I0026903

The
Children of
Cyclops

The Children of Cyclops

The Influences of Television Viewing on the Developing Human Brain

by Keith Buzzell

Waldorf
PUBLICATIONS
RESEARCH INSTITUTE FOR Waldorf EDUCATION

Printed with support from the Waldorf Curriculum Fund

Published by:

Waldorf Publications at the
Research Institute for Waldorf Education
38 Main Street
Chatham, NY 12037

Title: *The Children of Cyclops:*
The Influences of Television Viewing
on the Developing Human Brain
Author: Keith Buzzell
Editor: David Mitchell
Proofreader: Nancy Jane
Layout 2nd Edition 2015: Ann Erwin
Cover illustration: "3D TV static," licensed under public domain
via Wikimedia Commons

First edition ©1998 by AWSNA
Second edition © 2015 by Waldorf Publications
ISBN# 978-1-936367-86-3

Curriculum Series

The Publications Committee of the Research Institute is pleased to
bring forward this publication as part of its Curriculum Series. The
thoughts and ideas represented herein are solely those of the author
and do not necessarily represent any implied criteria set by Waldorf
Publications. It is our intention to stimulate as much writing and
thinking as possible about our curriculum, including diverse views.
Please contact us at patrice@waldorf-research.org with feedback on
this publication as well as requests for future work.

TABLE OF CONTENTS

INTRODUCTION

by Joseph Chilton Pearce

IN THIS UNIQUE, seminal, and disturbing study, Keith Buzzell shows that a major cause of our multifaceted social-ecological breakdown is a most common, near-universal practice followed by virtually everyone, worldwide, without a second thought. I have lectured and written on the subject of television for nearly thirty years and gathered a considerable amount of information on the subject, arguing from the beginning that the issue is not programming, but the instrument itself. In spite of all that, I felt like a babe in the woods when I read the stunning work which follows. Buzzell sheds so much light on, makes so much more comprehensible, all my own efforts that I must re-examine my entire approach to the subject. Nevertheless, with trepidations of carrying coals to Newcastle, I offer a few observations here as a preparation for the reader's challenge ahead.

For the following is challenging, and rightfully should be. Our current complacency, avoidance, and denial is our undoing. Buzzell faces the hurdle that general readers, victims of the very issue he addresses, are restless with other than thirty-second sound bytes, which can hardly suffice for a subject researched for ten years. Deep, pervasive, and potentially disastrous as this monumental juggernaut of destruction gone global is, television produces a mindset almost incapable of critical evaluation of what the device does to the mind.

As you will see, Keith Buzzell is well acquainted with the work of a long-time hero of mine, Paul MacLean, for nearly half a century the head of the National Institute of Mental Health (NIMH) research into brain evolution and behavior. MacLean showed why the mammalian species could have evolved only through a radically different form of reproduction, why gestation of a large-brained creature had to take place within the body of the egg-bearer, followed by a prolonged post-natal period of helplessness in the infant. There was no more laying one's eggs in the sand and forgetting them, as in our reptilian forebears.

A major reason for this is, to use contemporary metaphors, the difference between hard-ware and soft-ware. The more advanced an intelligence, the less its neural system can be "hard-wired." New evolutionary brains needed an imitative learning period during which "soft-ware" programs of ever increasing complexity could be constructed, a period for imprinting, following a model which would represent the latest update, on adaptation, including adaptation to changes wrought by previous adaptive strategies.

The greater the prospective intelligence and eventual adaptive ability, the longer the helpless period and the more critical the nature of the programming. We humans are the most versatile, adaptive creature of all life's productions—and our infants have not only the longest adaptive-learning period but the most profound helplessness known.

Paul MacLean spent decades tracing the essential requirements for the programming of such an infinitely open soft-ware as ours, and came up with a triad of needs resonant with the three different neural systems that make up our "triune brain," MacLean's most noteworthy observation. This "family triad" of survival needs is hard-wired into the cingulate-gyrus area added to the limbic structure of brain shared with all mammals. Simple, commonplace, and absolutely irreplaceable, this triad consists of nurturing, communication, and play. While the need is hard-wired into our brains, responses to these three behaviors can be met through nearly infinite variable means, essentially soft-ware. Though all mammals are hard-wired to provide this triad for their offspring, too, the materials for filling are more and more soft-ware, flexible and creative, as intelligence evolves.

A prolonged infancy of helplessness required that most memorable feature of mammalian life, those lovely mammary glands (that make dogs go mad and men leave home). Mammalian nurturing rests primarily on breast-feeding, and in humans the breast-heart-face and voice area is the indispensable matrix for new life, the safe-space in which the new mind can open to embrace its world, the primary source of not just the physical nurturing of body, but the foundation of communication and play, the nurturing of mind.

In the past century, however, medical interventions in the human birth process seriously disrupted the "species survival instincts" MacLean clearly noted, the "hard-wired" drives that compel a mother to provide this triad of nurturing, communication, and play, at all costs. While this nurturing drive is far more powerful in humans, it is still susceptible to disruption and dysfunction, as all higher intelligences are. Most serious of the broad range of disruptions wrought by technological birth was the elimination of breast feeding in 97% of the populace by mid-20th century, with overwhelming, globally catastrophic results.

A critical part of breast-feeding is the corresponding "in-arms period," the infant carrying or "baby-wearing" found throughout human history (until recently). This in-arms matrix assures and facilitates a cascade of critical developmental needs in the first post-natal year. Baby-wearing automatically provides the emotional, verbal, and pre-verbal communications around which a new world-self view is built in the infant. A mother's face at a distance of some 6 to 12 inches, as Fantz showed years ago, and her richly changing environment, provides the needed stimuli for building the visual world in the first year of life. As importantly, the in-arms matrix provides a prolonged and immediate proximity to the mother's mature electro-magnetic heart-field, and, as new studies show, the mother's heart-field literally lifts the infant's formative heart-field out of chaos into order. This stabilizes the infant's own heart function, which the new medical branch of neuro-cardiology (literally the brain-in-the-heart) finds is far more than just pumping blood.

Heart-brain interaction is profoundly important to neural growth, DNA, a stable immune system, and a raft of related growth issues. (Following our disruptive birth practices, both visual and heart dysfunctions have expanded exponentially in our populace, as well as breast cancer in non-nursing mothers, as Israeli doctors have pointed out. Meanwhile, our critical and unmet need for the breast then makes us equally susceptible to the constant use of breasts in advertising.)

So the disruptions in bonding at birth seriously undermine physical nurturing and the communication and play supposed to be established during that crucial in-arms period. Perhaps nature could have compensated for this violation of our ancient and foundational survival intelligences and the subsequent development hinging on them, but compensation itself began to break down when, within a decade of eliminating breast-feeding, television viewing arrived and literally possessed virtually 100% of the populace. The average American child spends some six thousand hours of TV viewing before age five, and Keith Buzzell has brilliantly explored the neurophysiological implications in this. He gives strong evidence that the act of viewing the device itself is the cause of trouble, not just the programming, a study that has not been done before. Above all, Buzzell asks the right questions for future research into the issue.

I would like to add that a critical developmental damage TV has brought about is in what it has taken the place of, substituted for, and/or prevented from taking place. First, TV disrupts or replaces verbal and non-verbal, emotional forms of communication between parent-child, already weakened by childbirth interventions and the lost intimacies of breast-feeding. Secondly, it has replaced story-telling, "grandmother tales," father's work-place accounts, and all the verbal chatter of the dinner table or fireside. It has replaced bedtime tales and turned radio from storyteller to music box. Thirdly, TV has replaced play. The child not played with does not learn to play, and play is the overarching intelligence of childhood and all learning (lifelong).

The drive for play is hard-wired; play as a capacity is pure soft-ware. Play is the primary way all learning takes place in the first decade of life (secondarily, lifelong) and intimately involves story-telling and/or family-talk, and the corresponding development of internal imaging which such word-flow fosters. Of all the damages wrought by TV, impairment of internal imagery may be the most serious.

All higher forms of intelligence on which a society depends, such as empathy, compassion, love, as well as the later stages of intellectual development, science, philosophy, religion, are based on capacities for abstract thought and the metaphoric-symbolic structures of mind developed through internal image-making, which begins formation in the first year of life. The foundations of the inner-image reside in the ancient mammalian, or "emotional-cognitive" brain and its interactions with the highest cortical, or "human," system, and should be well entrenched by the end of the first three years. Story-telling and family-talk are the cornerstone of this growth and disappeared in a majority of our populace when television appeared.

TV entertains the mind, and entertainment is not play. Entertainment cannot—by the nature of the formation of neural fields involved—educate, nor can TV. No "structures of knowledge," to use Piaget's term, can form, through shallow-dimensional sensory stimuli. This is one of the multitude of reasons that television has resulted in a steady incremental breakdown in educability in upwards of 70% of our population and every population buying into it. Television impairs or circumvents development of precise neural structures and mental abilities, and every country on earth importing our TV (and surprisingly few have escaped) and allowing (in fact fostering and encouraging) access to infants, toddlers, and children has undergone the same sequence of erosion of educational systems, cohesiveness of family and society, rise of crime and violence, child suicide, and general depression.

Along with the primary failure of infant-child nurturing, which follows in every country importing our technological birth practices, TV has compounded the breakdown. Michael Mendizza's *Touch the Future Foundation* newsletter has reported extensively on this. Ralph Nader and Linda Coco's recent book exposes a carefully researched commercial exploitation of the child mind, which I reported in my book *Evolutions End* in 1992. A *Forbes Magazine* article elaborated on how, with the appropriate programming, a child's buying habits can be set for life by the age of six, allowing management to set up long-range marketing strategies and production plans, making it well worth a multi-billion dollar corporate investment in the needed psychological research. The programming, of course, is through television, and the result is straight "pavlovist" conditioning. Nader quotes a major figure in the robustly lucrative MTV saying, "We don't influence the 14-year-olds, we own them." Ownership of the very early child-mind logically becomes the issue.

Communication in early developmental years involves, but is far from limited to, language or speech. Alexandria Luria's studies show how language serves far more fundamental biological needs of infant growth than communication in our adult sense. Coordinating the vast neural-spindle system, which saturates the body's muscular network and its connections with the cerebellum, is involved in language growth (as the work of Lewis Sander and William Condon of Boston University showed years ago). Keith Buzzell points out that the concentration of neural-spindles in the face, hands, and vocal apparatus is some ten to fifteen times greater than that in the rest of the body. It is not just coincidence that these are the very close and intimate associations provided by breast-feeding and amply concentrated on in close-up television. TV is designed to feed, in effect, on missing developmental stimuli, locking the viewer into visual addiction early in life, as we unconsciously search for the elusive nurturings long denied.

Language-muscular interaction is an intricate, delicate part of both play and development, and all play in the first seven years is verbal. No communication takes place through TV, however, where words and sounds are dramatically different than the rich, subtle, personal, and essentially quiet communicative environment given through baby-wearing and breast-feeding, as well as the long years of imaginative-imitative play the "child of the dream" is designed by nature to live in before the milk teeth come out and operational thought begins.

Language learning begins as a motor response in the last trimester of pregnancy and involves all those muscle spindles to build language into the entire body. It is well underway long before speech begins, speech being a different, more complex network. When hearing a story or family-talk, to which any hearing infant responds intelligently before speech begins, the brain forms an internal picture of any verbal description for which it has a neural pattern.

This effect rests on what is called the concrete language of the first seven years, when the same "sensory maps" concerning an object and the name of the object are a single set of neural fields. Speak the name and that single sound-stimuli activates the total sensory-map integration associated with that sound, engaging the entire brain in the creation of a facsimile of the original perception. Speak the word and the image appears in the "mind's eye," that inner-world picture which is the cornerstone of the human mind itself.

Creating an internal image in response to the stimuli of words is a major challenge to and involves virtually every facet of the brain, and it is the primary focus of development in the first seven years (and in modified form on through age eleven or so). Story-telling plays a preeminent role in this, as it has throughout history, and no society has been found that hasn't a rich repertoire of stories for the early child. The neurological effect is simple: The magical word comes in as primary stimulus, a frequency eliciting a whole-body response since *in utero*, and, in classical stimulus-response fashion, the brain constructs in response an internal image of the characters and actions the flow of words stimulates.

New neural connections in brain and body take place with each new story told, as new sensory-map connections form and integrate. Myelination of newly formed neural fields and the myriad sensory-maps involved permanently establish the growth and learning. Myelination of this sort requires repetition of the response generating the neural fields, which explains the child's love of and demand for incessant repetition of the same story. Within the second year of life, a child will begin to act out these multi-told tales through metaphoric play, where a material object "stands for" the internal image-object. This is the foundation for recognizing that symbols such as those of math or chemistry "stand for" universal functions and processes.

Television has not only replaced storytelling and family tabletalk, but in the pairing of audio and visual signals as a single sensory input, it has pre-empted the brain's critical need for the development of internal imagery as its response to language stimuli. The image is already there with the sound, the response pre-made with the stimuli, leaving the higher creative cortices with nothing to do. The vast fields involved in internal image making, the basis of all metaphoric-symbolic constructions and abstract thought, lie idle and eventually atrophy. Use it or lose it is nature's dictum, and we are losing it.

This particular impairment was demonstrated clearly when psychologists took a group of videos designed for children in the six- to seven-year age bracket and switched the sound tracks so that no sound matched the visual action. These garbled programs were played to a wide cross- section of children about the country, and with almost no exceptions, the children, no matter how carefully questioned, did not recognize the discrepancy. Their brains had habituated to a non- information source, precisely as Buzzell makes clear here, and had found the most economical neural routing for such meaningless sensory input, one that met no development needs and required no response.

The point is that the brain uses the same routing each time exposed to this paired audio-visual signal, whereas each new story told automatically expands neural structure. The programming on TV is, in this regard, almost irrelevant. Of the five to six thousand hours of TV viewing before age five, it may as well have been all one program. And at age five we force these young gods and goddesses into school desks, preventing bodily movement already denied by TV catatonia, and we demand from them responses to highly abstract metaphoric-symbolic systems—numbers and letters.

Forget in-depth comprehension of symbol systems such as alphabets and numbers from young people so deprived, or later comprehension of such awesome documents and ideals as the Bill of Rights. They won't have the neural structures for handling such, nor for generating compassion, love, empathy, understanding, care—those "higher human virtues without which a society destroys itself." These young people will have, however, as corporate advertisers have long known, a passion for consumption of material stuff they hope will alleviate the pieces of their life now missing.

Another pathology which will take on increased clarity as Buzzell's research unfolds, as mentioned above, is the fact that the brain habituates to a non-meaningful signal and relegates such a signal to the most rudimentary sensory processes. If nothing else is happening, the bulk of the brain literally goes to sleep; it has nothing to do. By the early 1960s it was found that young children particularly go essentially catatonic before the tube. Their eyes may be open, but nothing much is registering internally. TV producers found that by introducing "startle effects" such as sudden, unexpected noises or shifts of light intensity, the reticular activating system in the oldest survival brain sends an alarm which wakes up the higher cortical structures in the would-be-viewer, bringing attention to the source of the startle-effect. This wake-up call is brought about by the sympathetic nervous system giving the brain (and body) a brief shot of cortisol, one of the adrenal steroids that activates the brain-body in emergencies. Eventually, habituation to the early, benign forms of startle-effect took place, and the industry had to up the ante, to increase the intensity, extremity, and number of startle effects themselves.

Some twelve years ago Marcia Mikulac, of Brazil, published research that showed technological children's sensory systems to be 20 to 25% less aware of and/or responsive to environmental stimuli than children of preliterate or "primitive" societies. Recently, studies from the University of Tübingen and the Gesellschaft für Rationelle Psychologic in Munich, Germany, carried out in four thousand young people for twenty years, show that as a result of constant sensory over stimulation, a decrease of 1% per year in general sensitivity to stimuli has occurred, as well as a breakdown of neural ability to cross-index the various senses. This breakdown in neural synthesis leaves each sense a rather isolated sensation rather than part of the rich kinesthetic mixture of living events. More alarming, they found such desensitizing brings about a bypassing of the emotional, affective modes of the brain, those relational structures of mind that give the human being the sensitivities of love, compassion, appreciation, and so on.

The most serious of all indications, however, is that increasingly only ever-stronger, single-sense stimulation, titled the "brutal thrill" phenomenon, registers on the young people at all. Thus, we have the comment heard so often in the U.S.

that our young people are bored and apathetic toward ordinary living processes, restless and in need of ever-increasing external stimuli. Such a high-level input is found not only in heavy rock music, movies, TV and MTV, but in the ubiquitous internet (and computer viewing is a direct parallel to television in every aspect).

So, the desensitizing of the child's nervous system, brought about by failure of skin stimulus all nurturing mothers provide their infants (as Montague's and Marshall Klaus's studies showed decades ago) thus underwent dramatic worsening through television. The American child's need for more intense and ever-greater quantity of stimuli and corresponding insensitivity to most human values is an issue recognized by more and more people, while few are cognizant of cause.

While we as viewers grow blasé and sophisticated concerning the barrage of clashing image-sounds in the startle-effects now employed in both TV and movies, the ancient parts of the brain, those involved in activating the adrenals for the hard-wired defense postures through which we survived for eons of mammalian life, have no such discrimination. Our ancient survival systems react and release their shots of cortisol as hard-wired to do, even as our higher cortical "self-sense" expresses indifference. To the ancient, defensive parts of our brains, *the image is always true,"* as Buzzell puts it. This lower brain responds with its defense-alert even before we, as recipients, are aware of the image itself. (Between input of signal and our self-awareness of it, Buzzell estimates some forty million neural responses are inaugurated by the ancient survival systems of brain.)

As the rapidity and number of startle-effects increases, the parasympathetic nervous system has no time to bring balance back to the system, shut off the cortisol, reestablish the immune system, and so on. Thus, the body, as its own intelligence, lives in a constant state of threat or startle, and the brain suffers a serious over-saturation of cortisol. Cortisol, so vital to survival in the tiny emergency doses nature provides, is quite toxic in any quantity, and the symptom of cortisol overload is stress, proposed by the University of London's medical school as the *major cause* of contemporary disease, particularly cancer and heart trouble.

As serious as the corresponding diseases is the overproduction of neural connections from constant startle. At each startle effect and shot of cortisol, the neurons of the brain throw out new dendritic connecting links with other neurons, a redundancy that offers a wide scope of possibilities for fast, novel adaptation to emergency—a very ancient survival reaction of our brains that takes place before we are even aware of the event. When an emergency is over (environmental dangers are generally quite brief in mammalian life), this redundant neural mass is reduced back to normal size by the parasympathetic process, retaining only those new connections of neural fields involved in the "survival learning," if any, arising from that particular emergency.

If emergencies, or startles, follow each other in rapid fashion, however, as with the rapid shifts of major magnitude on current TV, the old brain has no choice but to keep the body's own system under constant hyper-alert, the redundant mass becomes more, and work will be ignored to our peril. Every cognizant, conscious parent or general reader should read and heed—indeed be nothing less than filled with righteous indignation and wrath, demand the follow-through research Buzzell calls for, and, above all, find the moral courage to throw their own diabolical TV device out the window. Now! Today is the day, and this is the hour!

PREFACE

An astounding and peculiar paradox characterizes the life of modern industrial man. One aspect of the paradox lies in the remarkable discoveries about the physical nature of our world and the resultant cornucopia of technological things that flow more and more rapidly from these fundamental insights. The other side of this paradox is the astonishing lack of intelligence demonstrated by man in the ways in which he introduces, indeed insists on his right-to-manifest, this effluent of things—without awareness of, or regard for, either the present context or the eventual consequences of these manifestations. The paradox is chillingly reminiscent of the leadership attitudes portrayed by Aldous Huxley in *Brave New World*.

With the push of a button a two-year-old child can now have a "controlling" influence in processes that required, for there to be a button to push, an extensive knowledge of quantum mechanics and relativity theory. Knowledge that a short time ago was accessible to a very few now underpins the button that activates a fractal elaboration, a multi-channel selector, or the real time accessibility of stock market transactions in Tokyo to someone living in Porter, Maine.

What is the significance of an infant's being able to set in motion forces and forms that penetrate and affect in specific ways the macromolecular- molecular- atomic- electronic and photonic worlds and bring about results that are both powerful and quasi-real? Is it the same as turning on a light switch or the ignition key of a car? Are life-like moving images treated by parts of our brain in the same way as naturally occurring sources of images? And if they are not, what difference does it make? Does a child's brain grow and develop the same when it takes in several hours a day of prefabricated images as it does when the physical, emotional, and intellectual parts of the brain are immersed in a varied "diet" of whole and real life-events? Does a high school graduate who has watched television for more hours than he/she has attended school differ in any brained capacity or potential from a similar high school graduate who has not? Has their physical or emotional life been affected in any verifiable way by 15–20,000 hours of television viewing? With the rapid expansion of interactive video and computer use by our children and grandchildren, their immersion in an ocean of prefabricated images will only increase. Does this have any effect on their growth and development, given that 600 million years of brained life grew and developed in a radically different life environment?

These and a host of similar questions pour forth when we begin to realize the profound changes in our life environment that have followed on the technological effluent of modern science's knowledge. Questions and challenges that life forms that have evolved over hundreds of millions of years *never* had to confront are now an every-day and constant circumstance.

This short book is an effort to begin to address a number of these types of questions, in particular as they relate to the human brain. It is short in large part because the inquiries into these questions have been few in number, and they barely scratch the surface of the biological processes that are affected by our headlong rush into the quantum-relativity world.

Hopefully, there will be many more books and research reports written in the immediate future by a spectrum of well-trained scientists.

Chapter I

The Children

WE ALL SHARE THE HOPE that our children and grandchildren will live in a world that is more harmonious, more enabling, more "whole." Our perception, shared by many of our generation, is that in today's world the fulfillment of that hope is, and will continue to be, exceedingly difficult.

In our search to understand why that fulfillment should be more difficult now than in prior times, we have concluded that three developmental features of the human brain have led to quite paradoxical vulnerabilities, and that these lie at the core of the difficulty. These vulnerabilities are biological in nature, appearing as a consequence of evolutionary challenges that were set in motion some 600 million years ago. They concern both the manner and the sequence of development of the brain's sensory-motor systems.

Stated in brief these features are:

1. The inherent vulnerability of the brain to the "images" it creates - or - "Which image is the real one?"

2. The hierarchy of neural processing times - or - "What do the body and the emotions react to before the mind finds out?"

3. The relativity of focus and context - or - "What is that giraffe doing in my living room?"

Neural Hierarchies and Processing Time

The brain can no longer be examined independently but must be viewed within its ecological context.
— Walsh,[1] 1980

Fundamental to our inquiry is a view of the structure/ function of the central and peripheral nervous system of the human being that we will summarize for later reference.

A. The central nervous system of the human being is hierarchically structured. At the apex of this hierarchy are those cerebral functions that mediate and/or are reflected in such distinctly human activities as thinking (e.g., analyzing, comparing, criticizing), speaking, reading, visualizing, and creating (each word taken in its broadest, unstructured sense). We ordinarily associate consciousness or awareness with these "humannesses," and considerable evidence indicates that the final expression of these functions requires multi-area and comprehensive, higher cortical activity.

B. Prior to the activation of these higher cortical regions, however, there is a complex mixture of interlocked, lower cortical and subcortical activities that, in turn, are temporally dependent on prior integrative processing and associative activities in such deep and ordinarily "unconscious" areas as the thalamus, hippocampus, and cerebellum.

C. Prior to in time, and yet "lower" in this hierarchy, are the senses and all of their sophisticated and subtle interrelationships: from their interface with the external world (or in the case of proprioception and balance, with the internal world), through their primary and secondary associative areas and the common integrative areas, to their presentation as "wholes" to the subcortical and midbrain centers.

D. These immensely complex levels of "digestion," which begin at the interface with the world and end at the cortex, occupy, in the time of neural integration, synaptic, associative and reverberative, a whole 0.5 second—the processing delay of Libet.[2] (Many levels of sensory integration and motor response do not require high cortical involvement.) Within this half second, however—and before we know it by cortical activation—a gamut of ever-widening streams of relationship are explored and completed. Synaptic relays can occupy as short an interval as 0.004 second. It is now known that a neuron can share information, in thousandths of a second, with at least 3–5,000 other neurons.[3] Throughout that half-second interval, the human being is not aware (has no coordinated higher cortical activation) of what is going on. What is going on is immensely subtle and far-reaching, however, as the energy transformed at the sensory interface moves upward through a hierarchy of centers on its way to the cortex.

E. It is evident that the vast majority, if not all, of these early patterns of integration-association are prewired; in any given moment of integrative processing, the routes, channels, etc., will follow essentially predetermined pathways. Such pathways have great history to them, extending far back into our mammalian past and beyond. In ordinary life they are not changeable by will or choice on our part. They are "prewired" to respond to certain energy inputs (be it light, sound, touch, etc.) and to share this transformed neural energy in complex pathways through hierarchically organized associative centers that have been a fixed part of the human being's nervous system for millions of years. (Prewired may be too simple an expression. There is great plasticity in neural differentiation, shown well in the progressive use of portions of the visual associative areas by the auditory cortex in persons born blind. A fundamental neuronal blueprint is present, however, from early embryonic development.) All of this occurs in this 0.5 second of "preconscious" time, before the cortex (or "we") are actively engaged.

The fact that our sensory-to-cortex patterns are essentially prewired and that millions of years of human physiologic development determine the way in which the matrix of energies that can be transformed at the external world interface will be responded to—these facts will be of considerable importance as we proceed.

Dependence and Vulnerability

From the 1960s to the present, Dr. Paul MacLean (Senior Research Scientist, Clinical Brain Disorders Branch, National Institute of Mental Health Neurosciences Center and author of *The Triune Brain in Evolution*) has been developing a consistent anatomical and neurophysiologic perspective that emphasizes the essentially "triune" (three in one) nature of the human brain. (See Figure 1)

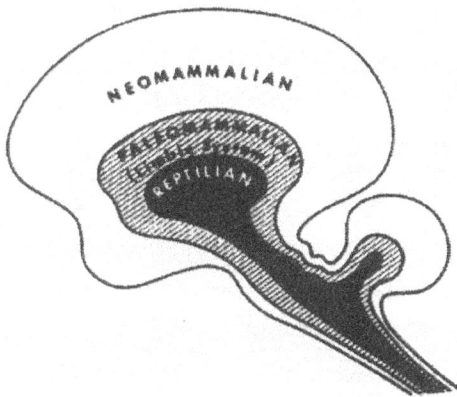

Figure 1

Fundamental to his view is that through evolutionary history, the central nervous system has progressively developed by unfolding three relatively distinct, "mentating instruments" or brains. While successively more complex "brained" creatures fully incorporate and refine the brain centers previously evolved, they also unfold new centers, with new potentials and capacities. These new centers interpenetrate the "older" centers, integrating and further refining their capacities. Within this essentially linear expansion and integrative growth, there are three qualitatively different foci or levels that give to the human brain (and other neomammalian species) a hierarchical organization that is becoming increasingly validated by research. The three hierarchical levels and their primary foci are as follows:

1. The Reptilian or Core Brain

MacLean refers to this as the R-complex, or the protoreptilian formation. It encompasses a number of ganglionic structures located at the base of the forebrain (the striatal complex) that are primarily concerned with "the regulation of the animal's daily master routine and subroutines (such as repetitious behaviors: emergence, basking, defecation, feeding, perching, etc.), as well as the behavioral manifestations of four main types of display used in prosematic communication (nonverbal signals: vocal, bodily, or chemical). The sensory-motor systems allied to the R-complex are the external senses (vision, hearing, taste, touch, smell) and the musculoskeletal apparatus. It could be said that core (or first) brain creatures are almost totally "hard-wired" and that their afferent-efferent (input-output) instruments are almost wholly dedicated to physical survival in the external world. MacLean terms the associative processing of the reptilian brain "protomentation." By this he is referring to "rudimentary mental processes underlying special and general forms of behavior including four basic forms of prosematic communication."

2. The Limbic Brain (paleomammalian formation)

afferent pathways

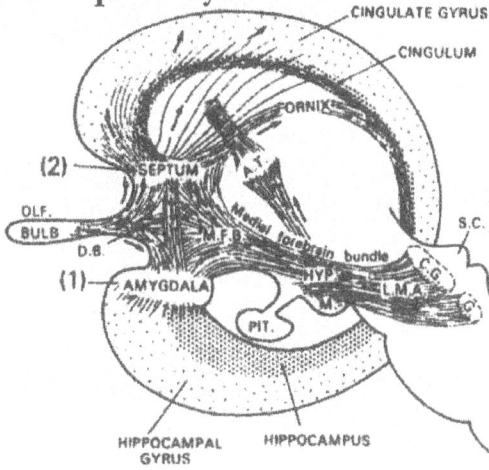

Figures 2a & b

efferent pathways

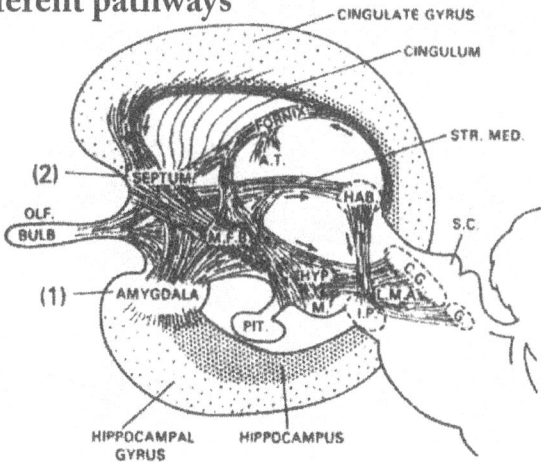

M.F.B. - *Medial forebrain bundle; PIT - Pituitary; Hyp - Hypothalamus; AT - Anterior Thalamus; M - Mammillary bodies; S.C. - Superior Colliculus; G. - Gudden's nucleus* [From MacLean, *The Triune Brain in Evolution*, with permission]

Mammals develop "three cardinal behaviors: (1) nursing-maternal care, (2) audio-vocal communication for maintaining and developing maternal-offspring contact, and (3) play." Taken together these developments mark the beginning of the evolution of the family. Underpinning these behaviors lies a long and elaborate development of the three portions of the limbic system: amygdala and septum, hippocampus, and the lamo-cingulate division. (See Figure 2)

A distinguishing characteristic of the limbic brain is its development of a large number of "internal" senses and motor instruments directed toward the monitoring of the dynamic internal metabolic states and the "wholistic" expression of those states (i.e., externally via gesture, posture, tone of voice and facial expression; internally via shunting of blood supply and spindle control of skeletal muscle states of contraction). Additionally, as recent research has explosively demonstrated, the limbic brain is bathed in a wide variety of neuropeptide-transmitter-hormonal chemicals that link emotional states and the physical-physiological and immune systems. (See Candice Pert, "Neuropeptides: The Emotions and Bodymind," *Noetic Sciences Review*, #2 1987)

As noted previously the core brain is both enlarged and enhanced throughout the two hundred million year history of the limbic brain. While intimately integrated in the "wholeness" function of a mammal, the core and limbic brains retain a significant degree of autonomous function. The first brain still has its primary focus on the external world and physical survival. The limbic (second) brain, receiving massive interoceptive (interior) sensory input, has a primary focus on the interior world of dynamic states, reflected in the emotions, the emergence of MacLean's "Sense-of-Self," and the subtle but enormously influential modification of the musculoskeletal expression of affect (emotion) via the muscle spindle and Golgi tendon system. The external secretion of some of the products of the limbic neuropeptide pool further enhance that expression. Survival, in a limbic brain context, appears to focus on the preservation/ enhancement of the creature's hierarchal place within the emerging family grouping. MacLean calls this form of cerebral activity "emotional mentation, a form of cerebration that appears to influence behavior on the basis of information subjectively manifest as emotional feelings."

3. The Neomammalian Brain

The development of the high cortices (see Figure 1) characterizes the emergence and hierarchal separation of the third, or neomammalian brain. While the high primates come to a formidable development of certain cortical centers, the third brain is not considered as wholly unfolded until the appearance of the human being.

As with the limbic brain, the neomammalian brain develops unique centers and deeply interpenetrates (and is penetrated by) the limbic and core brains, interweaving all the basic functions into a responsive "whole" (hence triune—one brain but with three aspects: the physical, emotional, and intellectual).

The third brain receives extensive input from the limbic and core brains. (For reasons not yet clear, the core brain appears to supply greater input than the limbic.) This dual input places the third brain in a unique circumstance. Quite aside from the input deriving from the other two sources, the third brain has the unique capacity for rational mentation (ratiocination). Included under this very general heading are such distinguishing human capacities as the ability to formulate questions, inquire, analyze, see abstract relationships, work with numbers, speak, write, invent, and artistically create.

The neomammalian brain establishes a "third" viewing place, relatively able to be separated from the exterior and interior "worlds" of the core and limbic brains and, while drawing on their input and images, conducts a cerebration process that is distinctly different from either emotional mentation or protomentation. When functioning in this hierarchically separated state, the third brain is quite differently oriented relative to the question of survival. By being able to incorporate the other two brains' input but remain, to a degree, separated from both, the third brain can focus beyond physical or self-group "survival" and concern itself with the "survival" of its functional results, namely, the ideas, understandings, concepts, and creations that flow from its capacity to abstract and give new "forms" to meaning. Examples of these unique third brain coalesced capacities are legion but are amply illustrated by the lives of such persons as Pythagoras, Newton, Bach, daVinci, St. Theresa of Avila, and Moses.

The foregoing sketch of the triune brain of the human being establishes an evolutionary hierarchy which continues to mirror this threefold emergence in the manner in which the outside world (1) enters via the external senses (core brain), (2) moves up into the limbic brain, and then (3) enters the mammalian (neocortical) brain. In other words, the neocortex is dependent on sensory data deriving from the five senses, and this data is, in temporal terms, initially processed by the mentating instruments of the first (core) brain. The images that result from this processing then move upward (relatively) into and through the limbic brain before they are available to the neocortex. The significance, or evolutionary survival (neomammalian brain) meaning, of these images will begin to evoke responses from neural centers lying within the core and limbic brains before the image data is fully processed by the neomammalian brain.

For the most part this is a very fortunate arrangement as it makes possible many rapid survival responses (like ducking one's head or jumping aside to avoid a rapidly moving object or averting the eyes from intense light). Humans react to these potentially harmful influences long before (in neurological time) the neomammalian brain can process its level of the meaning of the image. Similarly, humans experience emotional reactions (e.g., "I was angry before I knew it") before becoming aware of the limbic brain's evocation of a range of responses (e.g., muscular tension, rapid heartbeat and respiration). Also experienced is the more prolonged emotional reactivity (the results of neuropeptide release) that follows a near accident or fright. Often this emotional reactivity persists even when the neomammalian brain becomes aware of the unreality or innocence of whatever provoked the emotional reaction.

The disparity in processing times, via evolutionary necessity, creates one order of sensitivity/vulnerability. This can be capsuled by saying that humans often react physically and emotionally before any neomammalian coalescence of meaning (the more abstract images of the third brain) can occur.

The dependence of the neocortex on the limbic and core brains shows itself even more dramatically when we consider the vulnerability of each of the brains to the images or resonant representations it constructs and shares. The immense survival value which accrues from being able to form resonant representations or images of the outside and inside worlds has a high price attached to it. The price is that the images must be taken as real. For as long as one-, two-, and three-brained

31

creatures lived in the natural world, the requirement that images be taken as reality served the life form extraordinarily well. When early man appeared, however, the use of deception for its survival advantage (present in elemental form even in reptiles) quickly underwent a broad diversification. The decoy contains the essence of this "deception" in image. As time passed and early communities formed, there is extensive evidence of the use of other prefabricated or synthetic images, of "look-a-likes" and "are-a-likes," which serve not only the physical survival thrust of the first (core) brain of the human being but also the self-group survival thrust of the second (limbic) brain. One dramatic illustration of this is reflected in dress and in the wearing of tokens or emblems. The hierarchal ordering of early communities was partially defined by the formalized use of these prefabricated images. Soon (in human evolutionary time) the use of synthetic images became applied to other aspects of the social and religious life of communities, often with very beneficial results (images of higher powers, of spiritual and social leaders and of human communication, values and creative expressions).

With the further development of human societies, there came the inevitable use of prefabricated images that served the aims of a more rigid social hierarchy (master-slave or kings and subjects, etc.). With the appearance of large-scale trade and of monetary systems, the use of these images for advertising purposes began. Within the past two centuries there has been such a crescendo of "image-for-advertising" that we can scarcely imagine a world without it.

All of these developments derive, ultimately, from the built-in vulnerability of each of the three brains to their corresponding images. From the pinnacle of brained survival mechanisms, the hard-wired *belief in* image has exposed a vulnerability of unmeasured dimensions. The arrival of television exposed this vulnerability to depths that are still being plumbed by the TV industry.

Images

In everyday use the word *image* is most often applied in the visual sense of the term. In this book we will be using *image* in its broader meaning wherein it refers to any proportionate or resonant representation. Hence, we will be referring to the products of our hearing center (auditory cortex) as sound images; the product of our tactile cortex as touch images, etc. In truth, each of our sensory systems constructs or creates a proportionate representation—an image—of that aspect of the outside or inside world to which it is sensitive.

In the case of our olfactory (smell) cortex, this is less obviously an image for us humans. We have only 6 million olfactory cells in our nose. A wolf, with 270 million olfactory cells, images the molecular-in-air world in quite remarkable detail. Similarly, but by making use of ultrasonic waves, the bat brain constructs a resonant representation or image of the world it moves through. Each of our five external senses and their associative centers construct a resonant representation or image of that portion of the world with which its sensory end organs interface. The brain then fuses these images together into a resonant representation of the world around us.

Although not usually spoken of in this way, it is also consistent to refer to the coalesced products of corresponding associative centers in the limbic (emotional) and neomammalian (intellectual) brain as images or resonant (proportionate) representations. For example, the gamma system (muscle spindle and Golgi tendon organ) underlies all of the subtle expressiveness of the face, hands, neck, and vocal apparatus. Via the coalescence of its input with other input to the limbic brain, an image or resonant representation of our inner emotional state is created and portrayed in facial expression, gesture, posture, and tone of voice.

In turn the neuropeptide pool, generated primarily by activity of the limbic brain, creates a molecular image of our affective state that contributes greatly to the subjective experience of feelings or emotions. Further, MacLean in his book *The Triune Brain in Evolution*, speaks of the coalesced input of the external (exteroceptive) and internal (interoceptive or emotional) senses as being the data from which the image he calls "the Sense of Serf" is created. What modern psychology refers to as "self-image" is another resonant representation (a subjectively experienced one) formed from the continuity of one's experience (memory), the state of the neuropeptide pool, and the coalesced external and internal sensory data.

Neomammalian (neocortex) brain images, while considerably more abstracted, can be understood, nonetheless, as resonant representations. Vowel and consonant sounds (micro-images) are combined into words, and these into sentences. Each of these is a resonant representation—an "image that stands for" increasingly complex wholes. We weave pictures from words in poetry, novels, and our own descriptions of the world around and within us. We create images—in geometric forms, architectural drawings, and artistic paintings—even in dreams. With number (another order of resonant representation), we abstract and compound image even more—and come to arithmetic and mathematics.

Many of our common expressions fuse these three "orders-of-image" together into one, subtle whole. For example, a man of my acquaintance was referred to as having a "glacial exterior" (the comment came after several days of interacting with him). The massive, cold bulk of a glacier is clearly an external sense image. This is melded into a subjective (or feeling) whole by the relative resonance between a real glacier and certain personality features of the man. He appeared "cold," unspontaneous and "heavy," with a bluntness that changed slowly if at all. His facial expressions, gestures, and tone of voice were resonant with both the exterior image of the glacier and the subjective image of his being unemotional, colorless, and lacking warmth. Only the neomammalian brain can fuse these two sources of image and express the whole impression in words (the third, more abstracted order-of-image).

Similarly, the Statue of Liberty is a coalescence of all three orders, or levels, of resonant representation. The massive metallic bulk, the female form, the crown and torch—these are all external (visual and touch) images. The subjective, emotional images (of the sense or feelings of security, welcome, pride, strength, etc.) arise from the subjective world of the "Sense of Self" (limbic brain)—a coalescence of the external and internal. The more abstracted neomammalian images evoked are captured by words and expressions like: liberty, opportunity, international friendship, and asylum.

In summary, the human brain, via its three levels of sensory input and associative centers, constructs moving images (or resonant representations) that fall into three diverse but consistent categories:

1. *Essentially non-abstract images of the external world.* The sensory systems of the reptilian (core) brain create quite precise, resonant, but relatively concrete reproductions of a portion of the forms and energies lying exterior to the body.

2. *Subjective images of feeling, or emotional states.* These images are blendings of the more concrete external with the more subtle internal. Because they are blendings, these images lie along a pendulum swing of subjective states, beginning with a reflection of bodily sensations (e.g., as when we are in pain, we answer, with irritation, "Because it hurts!") and extending to a deeply felt compassion for the suffering of another person. A simplified illustration of this gamut of feeling states would be this:

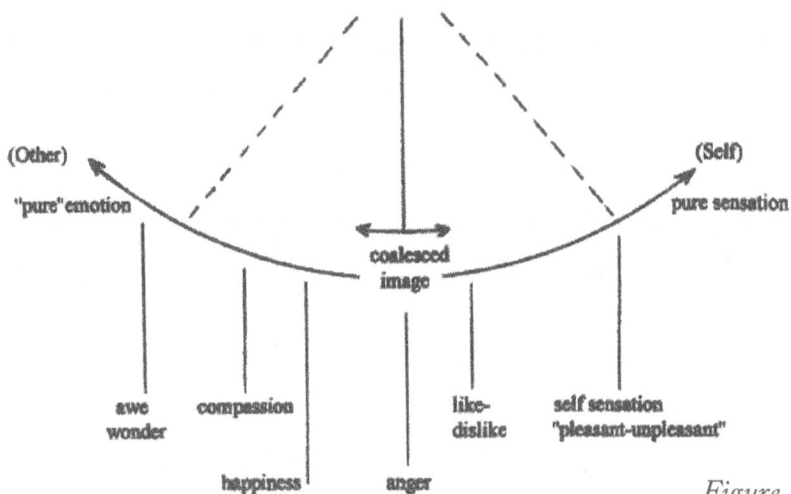

Figure 3

A further qualification of affective (emotional) images derives from their variable persistence in time. The images that depend on and, in turn, evoke large scale neuropeptide release tend to prolong the subjective experience. Memories, coalesced life values, and the variable participation of the neomammalian brain also serve to modulate the temporal quality of these images. An order, or level, of subjectivization is entered with this type of image. It is "my anger," "your coldness," or "our enthusiasm" which is experienced inside, although it is usually reflected outside as well (through facial expression, gesture, and tone of voice). The images are, to a degree, abstract (following the definition of abstract as "considered apart from matter or the concrete").

3. *Abstract images constructed by neomammalian (neocortical) associative centers.* Language, writing, arithmetic, geometry and mathematics, analogical thought, spatial visualization, musical notation—all these capacities are grounded in image formation, in building abstracted but resonant representations. The images are now "symbols-that-stand-for" and become, over human evolutionary time, more and more abstracted and interwoven. The concrete and subjective images of the first (reptilian or core) and second (limbic) brains form much of the elemental "food" for these remarkable and progressive abstracting abilities of the associative high cortex.

For as long as brained life evolved in a natural world, a world not of its making and thus containing only naturally occurring sources of image, these three developmental features operated to the distinct advantage of the brained creature. Shortly after the human being appeared, however, he began to make utilitarian, artistic, religious, and communicative use of prefabricated or synthetic images, and with this the brain's inherent vulnerability to image showed itself. (Such words as creative, pre-constructed, non-natural, imitative, and fictitious could also be applied to these images, but each carries connotations that are incomplete or not always appropriate.)

By prefabricated or synthetic we refer to the making or creating, by man, of a host of different sources-of-image— from hunting decoys, early cave drawings, carvings on tools, sculptures, tokens, etc., all the way to oral and written language and symbols. The second and third features (neural processing times and focus and context), also present from the beginning evolution of brained systems, combine with the first feature (vulnerability to image) to add to the hazards of living in a world where naturally occurring and synthetic images became more difficult or impossible to tell apart.

The use of prefabricated images increased incrementally throughout humanity's early social and cultural history. From the time of the Renaissance (with the invention of the printing press and other technological expressions deriving from science) until well into the 20th century, prefabricated images played a progressively larger role in the physical, emotional, and intellectual life of humankind. In particular I am referring to radio, movies, print, advertising, etc. In the immediate post-World War II era, with the first technological applications of quantum mechanics and relativity theory (e.g., the transistor, maser, laser, television, and early computers), the use of prefabricated images by all aspects of Western society began an accelerated growth that resembles the rocket trail of a Saturn booster. The majority of these uses were and are finely honed to focus on the inherent image vulnerability of the human brain. Now into the late 1990s we live immersed in an ocean of these prefabricated images, almost totally unaware of their real and potential neuro-biological consequences.

It is our aim to explore a number of those consequences, and we have chosen television viewing as the optimum means to facilitate that exploration because it is clearly the most potent and the most widespread application of prefabricated images in the modern world. In addition it is the single most time-consuming event, with the exception of sleep, in which a majority of American children participate.

Our concern throughout the exploration will be with television viewing as a physical-neurological event. There will, therefore, be no consideration given to program content itself. In a sense we will be looking at the "digestion" of a television image in a way similar to the study of the physiology of food digestion, rather than study the varieties of food that could be eaten. The emphasis, therefore, will be on the mechanics, pathways, and steps in the process by which the brain digests a television signal.

We must also make note of the small number of published research reports on television viewing as a bio-neurological event. The contentious and voluminous literature focusing on program content has, to this point in time, almost entirely occupied the attention of the public and of the research community. The present availability of sophisticated research tools, however, removes the final impediment to a careful study and elucidation of the bio-neurological consequences of television viewing. Because those consequences are intimately involved in all stages and processes of brain development, it is of particular importance that we come to a much clearer understanding of their role in the normal growth and development of our children. Notwithstanding this, the three features and their consequences apply as well to the adult brain.

In the final chapters we will consider a number of the implications that flow from the research findings reported to date. The discussion of those implications will make indelibly clear the great difficulties referred to in the opening paragraph of the Preface.

The Relativity of Focus and Context

The human being's visual system is an excellent example of the third mentioned feature: focus and context. The central part of the retina, called the fovea or macula, has a much higher concentration of cones (the color sensitive cells that transform the visible light spectrum into neural impulses) than the periphery of the retina. (See Figure 4)

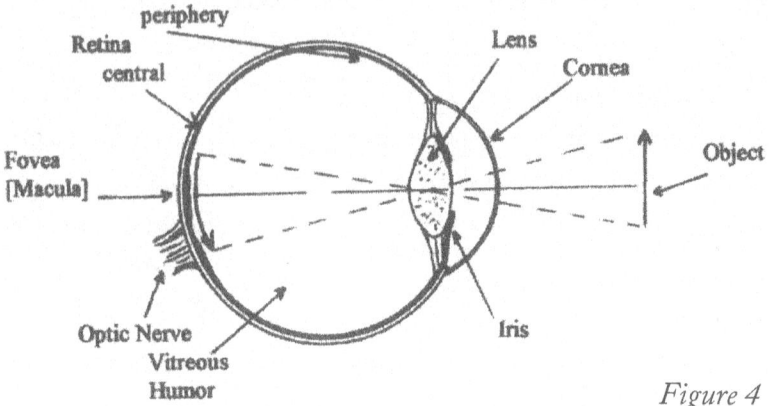

Figure 4

This makes it possible for the lens to focus the light reflected from objects onto the most sensitive, central part of the retina. The visual association areas, therefore, receive more data from the macula and can construct a clearer image than is possible from the data received from the periphery. What we often do not appreciate, however, is that our peripheral vision provides the background, or context, for whatever we are focused on. If we did not have peripheral vision to provide this context, then the meaning or significance of the event would be greatly reduced or absent.

The ability to focus and to place what is focused upon into the real background of that moment is one of the most essential and remarkable capacities of the brain. We can experience this remarkableness every time we slowly scan the horizon, search out the ripe blueberries, or notice when our smiling child is standing too close to the edge of a step. This capacity is also evident when we listen carefully to the expression of another's sorrow and are aware of how important it is to place this sorrow into a right context of the moment and of the other's whole life. The comparative-analytical part of our third brain demonstrates this simultaneity on each occasion that we test a hypothesis, try to integrate new research findings into our established background of understandings, or try to find just the right word to highlight the expression of a complex idea.

Exactly what makes it possible for us to keep, narrow, move, or enlarge our focus, and blend, enlarge, or contract the context is not well understood as yet. What is known is that it requires widespread involvement of the reticular activating system (the germinal source of attention and vigilance mechanisms) as well as multiple, near-simultaneous, and active participation by multiple subcortical and cortical centers. The capacity to refine, vary, and harmoniously blend that which is focused upon with the total milieu or context could, thus, be taken as a benchmark of intelligence.

This intelligence is evident in the world of the cold-blooded reptile, as when one observes a lizard searching for food—comparing, evaluating, selecting—all within the spatial background of its territory. It is evident, but far greater in its breadth of capacity, in warm-blooded mammalian life, seen in the selective wholistic behavior of a mother for its litter, feeding and nurturing each (focus) and simultaneously being aware of the others, the time of day, the indicators of satiety and of possible danger (context). It reaches its acme (potentially) in the triune brain of man where the points of focus and the breadth and nuance of context can encompass such notions as the singular value of each human being within a context of 5.5 billion people and in the search for elegantly simple theories (focus) that encompass infinitely diverse functional manifestations (e.g., $E=MC^2$ or the triad of family behavior patterns flowing from the cingulate gyrus of the brain).

The progressive refinement of each of the sensory systems of vertebrate life was accompanied by an inner qualitative distinction between the specific and the general, referred to here as focus and context, or singularity within the whole. Approached from this perspective, we could say that all deepenings of meaning and understanding are the result of the increasingly harmonious fusion of what is focused upon (the singularity) with the "arena" or background in which the singularity is functionally embedded.

Each of our external and internal sensory systems can be brought under the wholing influence of this harmonious fusion of focus and context. To the degree that we are increasingly able to accomplish this within the changing flow of our physical, emotional, and intellectual life, we are aware of more aspects (focus) of reality and how they can be appropriately blended together (context); hence, we are more intelligent.

From the time when prefabricated or synthetic images (a manifested focus) began to be introduced by human beings, the likelihood of the brain's accepting a synthetic image on the same basis as a naturally occurring one increased incrementally. This can be a harmonious and creative influence—as witnessed by the synthetic but highly resonant and intelligent images that underpin language, poetry and myth, painting, music-making, geometry, and mathematics.

What is essential to underscore here is that the sensory and associative aspects of the physical (core or reptilian) and emotional (limbic) brains cannot, of themselves, tell the difference between a real and a carefully crafted synthetic image. (This is dramatically illustrated by the ingestion of countless, harmful "images" by cold- and warm-blooded creatures. The image, whether plastic, metal, or wood, is *mistaken* by the creature for a naturally occurring source of image. Our core and limbic brains are very often misled by the same unreasoned similarity.) They will process to their maximum capacity whatever is presented to the sensory interfaces and will initiate motor responses consistent with the meaning or significance of that data. The meaning itself is relative to the associative fusion process carried out by each, or a combination of, the triune aspects of the brain. For the physical or reptilian brain, meaning is always reconciled by the physical survival imperative; significance for the limbic brain is reconciled by the survival imperative that relates to self-group hierarchal status (Sense of Self and Self-image).

Here again we must emphasize that the core and limbic brains cannot reason; they cannot analyze, critique, or evaluate independently of their sensory and associative cortical activity. Only the third, or neomammalian, cortex has the associative-discriminating power to do that, and then only if it is fully and resonantly engaged. If the neomammalian brain is inhibited or disharmonized by the sources of synthetic images, then it cannot reason, cannot weigh, discriminate, evaluate, or wholly participate in the necessary fusion of the focus and context (the meaning) deriving from the associative activity of the core and limbic brains or its own multi-leveled cortex. This is the focal concern, the central question deriving from the results of the limited research done to date on television viewing as a neurophysiologic event. There is evidence of disharmonious and fracturing influences of television viewing on essential brain mechanisms at a number of levels. While this evidence is not conclusive or detailed at this time, its presence cries out for impartial multi-disciplinary evaluation, research, and clarification. We should accept no less, for the real good of our children and grandchildren.

The Growing Brain

Research on the growth and development of the brain is currently one of the most fermentative fields in neurobiology. It would be impossible to summarize all that has been tentatively or conclusively established over the past twenty years, but it is important, relative to our aim, to note the essence of a number of studies.

Particularly pertinent to our thesis is the observation that there are developmental windows in the brain. Recent research has established that there is a time scale during which maximal neuronal interconnectedness occurs, and that these windows of opportunity vary with different capacities. In several instances the window appears to close at the end of its cycle of development and is exceedingly difficult or impossible to open again. Others remain open to varying degrees but require great effort to establish new connections. The following is a brief summary of the known windows and their relative sensitivity to time.

Overall brain development requires a rich and varied diet of stimulating experiences. This begins very early, and in children who play little and are not nurtured by a great deal of cuddling, touching, and being talked to, the brain may be reduced in overall size by as much as 25–30 %.

2. The primal social-relational development of the child is relatively complete by age three. The brains of children who are abused or neglected during the first three years of life undergo both mis-connections and a lack of rich axonal-dendritic interconnections between essential frontal and limbic centers. There is clear evidence that these are extremely difficult or impossible to modify after the age of three.

3. Visual acuity and full binocular vision are fully developed by age four. Constant and varied visual stimulation is essential from birth on.

4. Complex feelings like joy, envy, and empathy require longer to complete their wiring, the process continuing until age nine or ten.

5. Language development is phased and interdependent (i.e., earliest are vowel sounds, later come combinations of sounds, words, simple sentences, etc.) Speech recognition is completed around age seven to eight but continues, with less facility, throughout life.

6. Basic motor skills (upright stance, walking, hand coordination) develop with remarkable speed during the first four years. Fine motor skills begin at about the fourth year and begin to wane in rapidity of absorption by age ten.

7. The rapid phase of brain development slows at around age ten. Minimally used synaptic junctions are dropped in an accelerated burst of the final coalescence of pathways that have been conditioned by repetitious experience. The earlier overproduction of synaptic connections establishes an incredibly broad context of possibilities. Real life experience then hones this plenum of possibilities much like a sculptor chips away the stone that is unnecessary to the refined creation.

Throughout the evolutionary time of the human being (until very recently), each step in this long but rapid process has been fed by whole, real, three-dimensional (physical, emotional, intellectual) experiences. As recent research has confirmed, a brain cannot develop harmoniously without a rich and varied mixture of them. It is here that special emphasis must be placed on the role of technologically derived synthetic or prefabricated images. If the sources of image (that which is focused upon), and the life-event (context) surrounding the introduction of those images, do not present a balanced whole which can be experienced by the child in the real time of the event, then a type of split or disharmonized brain circumstance results. Learning, when defined as an activity which should engage the resonant physical, emotional, and intellectual aspects of our body-brain, is a fantasy under these circumstances.

Throughout the 2.5 million years of human presence on the planet, children grew through their developmental years immersed in whole events. These events, however repetitious and dull, dramatic or life-threatening, were real. When adrenaline was released, heart and respiratory rate increased, and blood flow into muscles surged; there was a real event going on in the outside world that woke up, alerted, and involved the whole of that young person: their body, their emotions, and their mind. When we consider the newly emerging data on developmental windows, on the rapidity of neuronal growth, and the rapid closure and elimination of associative cortical connections at around ten years of age, and place these rigid biological requirements into the context of 20,000 hours of unwholed, repetitious, emotionally evocative but physically unfulfilled experiences, we can gain a bit of perspective on the dimensions of the difficulty.

Chapter II

The Neurophysiology of Television Viewing

It is instructive to consider that the word health *in English is based on an Anglo-Saxon word* hole *meaning "whole"; that is, to be healthy is to be whole.*
— David Bohm[4]

DURING THE PAST FORTY-FIVE YEARS billions of words have been written and spoken on the positive and/or negative aspects of television programming. Considering the ubiquity of television in the Western world, this volume of commentary, evaluation, and opinion is not surprising. What is surprising, in fact what should be quite startling, is that so few questions have been asked and so little research has been done which are directly related to the question of television viewing's effects on human physiology (biochemical, endocrine, neuromuscular-sensory, and central nervous system processes).

It has been a commonly held point of view that television is simply one more media and that it is inherently, as a neurophysiologic or brained event, the same as its predecessors: print, radio, recording, and movies. A primary purpose of this book is to establish that this presumption is incorrect and misleading in a number of ways. To accomplish this, our focus throughout will be on physical and neural processes that are affected by television viewing regardless of program content.

Aims:

1. To establish the legitimacy and urgency of a multifaceted research program focusing on the physiologic responses of humans to television viewing.

2. To discuss a number of dependencies and resultant vulnerabilities of the human brain. These will be explored in the context of the brain's triune evolutionary unfolding, the times of neural processing, and the extraordinary sensitivity of the brain to image.

3. To review the scientific literature pertinent to the known neurosensory responses of humans, primarily to visual stimuli. Particular emphasis will be given here to those processes that are now understood to directly reflect, both qualitatively and quantitatively, such functions as attention, vigilance, and higher cerebral activation (as indicators of thought, analysis, criticism, evaluation, etc.). This will establish a series of comparators to be used in studying the responses of humans to television viewing.

4. To discuss the profound differences between radiant and reflected (ambient) light in the physical and structural sense and to show how their differences present totally different electromagnetic environments to which humans respond in quite different ways.

5. To review and discuss the known but limited research literature dealing with neurophysiologic responses to television viewing, including the role of neural habituation.

6. To review a number of recent studies which focus on the physical (biochemical and whole body) effects of television viewing.

7. To discuss the phenomenon of television epilepsy, its causes insofar as they are known, and the implications of its presence as a neurophysiologic abnormality.

8. To present and discuss the implications of television viewing as an activity in which the viewer is subjected to a form of multi-leveled sensory deprivation.

9. To outline arenas of research that could clarify the many unanswered questions.

The following self-evident givens are assumed without further discussion.

A. It is possible to describe and delineate a typical environmental situation in which the vast majority of television viewing takes place.

B. A functioning television set is a technologic device with features that can clearly be defined as essential to its operation without reference to the program being televised.

C. When a television set is being watched, numerous and varied physiologic reactions and responses take place in the viewer. Many of these are measurable in a variety of non-invasive ways.

A Significant Subject for Research?

Over the past forty years a wide variety of human activities has been subjected to intense, prolonged, and sophisticated research, e.g., eating, sleeping, varied sports or physical activities, sexual intercourse, aspects of working, classroom learning. Thousands of books, research reports, evaluation and opinions, and widespread discussions up to and including large scale political action have resulted from directed attention to the physiological significance of these human activities.

With respect to television viewing as a neurologic-physiological event, however, remarkably few studies have been undertaken, and even these have been quite limited in scope, number of subjects tested, and follow-up. They are so few in number that only a small community of researchers and writers has undertaken to comment on them. The near universal response from the medical, scientific, and lay public communities has been protracted silence and apparent indifference.

The Nielsen Media Research Corp. is still the only industry-wide source of data which reflects TV viewing habits. While its data has been criticized by a number of commentators (see Horst Stipp's *American Demographics Magazine* from March 1997), it is reasonable to take four hours per day (more modest than Nielsen's data) as the time the average American spends watching TV. Given that there are approximately 97 million TV households, 39 million children aged 2–11, and that video usage is not included in the estimates, it is fair to say that the average American watches at least 1500 hours of TV per year, and that children will spend considerably more hours watching TV than attending school, K–12.

Taking the most conservative viewing figures available, it is a simple matter to calculate that the population of the United States alone is involved in an activity that yearly consumes in excess of 350 billion hours. As noted before, the ubiquity of television throughout the world makes this observation of total hours consumed in the United States a very partial reflection of the total time consumed by modern man. Added to these figures must be those deriving from the growth of the video and computer industry. Nielsen reports that maximum recording of network programming was on Saturday and Sunday prime time.[5] These are the same days the maximum use of rented tapes occurs. While viewing home videos does displace a variable portion of commercial programming, there is evidence that it has lengthened the total hours cited by Nielsen.[6] We must also cite the mushrooming of video use for a large number of educational and job training purposes.

As impressive as these gross numbers are, we must note that none of the previously mentioned human activities (except sleep) that have received concerted research effort even begins to approach a significant fraction of television viewing time. These comparative figures highlight an extraordinary and paradoxical situation. It is more than a bit parallel to other 20th century technological/industrial developments (e.g., from radium watches, the irradiation of children's thymus glands, to DDT) that were presumed to be harmless, and with respect to which it took many years before fundamental research uncovered their multiple hazards. The headlong rush into utility and profits that has characterized much of the history of Western applied science has a particularly emphatic example in the essentially unquestioned and unresearched physiological effects of the most rapidly expanding and powerful technology (television) that has ever appeared.

An activity in which hundreds of billions of hours are invested by human beings each year would, on a quantitative basis alone, appear to be a worthy research subject. As we shall endeavor to establish, there are a host of compelling qualitative reasons why a call for an urgent and broad-based research effort into the physiological responses of human beings to television viewing is appropriate.

Visual Response – Cortical Activation

*Perceptual systems are exploratory. In the absence of
adequate information the perceptual system hunts. It
tries to find meaning, to make sense from what little
information it can get.*   – J.J. Gibson

In the human being a number of responses of the visual
system are primary indicators of cortical activation.[7,8] While many
of the cerebral networks activated during visual systems arousal
lie below what we ordinarily consider to be conscious levels, they,
nonetheless, reflect intense and widespread mid- and higher-
brain processes that involve attention and vigilance mechanisms
that are essential for higher cortical functioning (e.g., analysis,
criticism, comparison—learning in the characteristically human
sense).

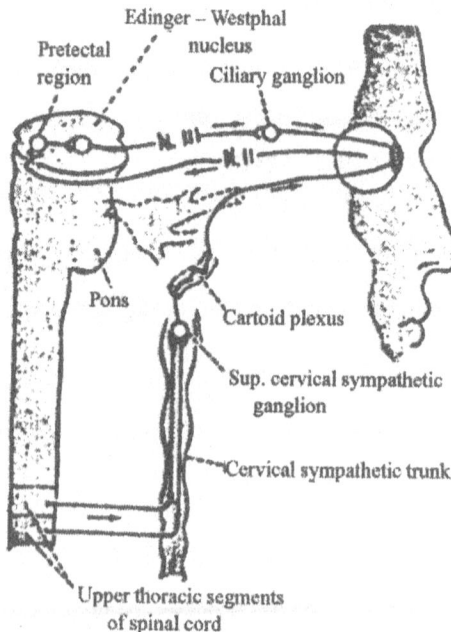

Figure 5a

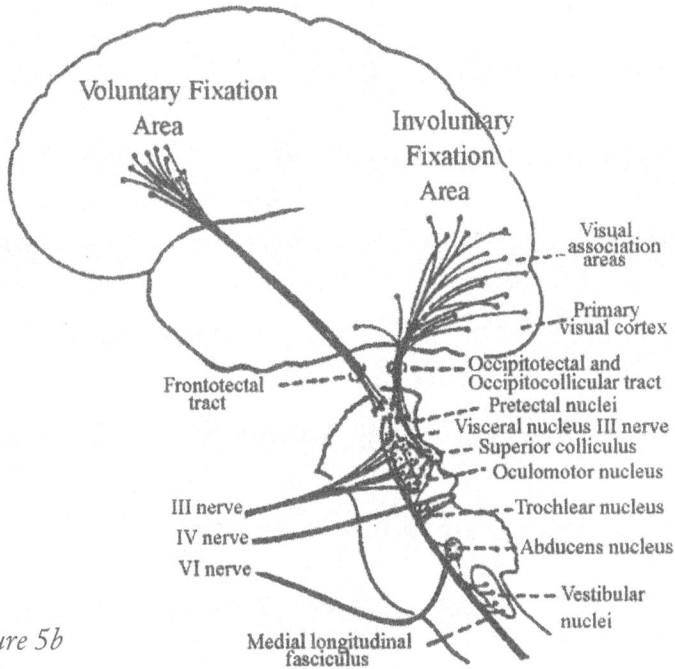

Figure 5b

Voluntary Fixation Area

Involuntary Fixation Area

Visual association areas

Primary visual cortex

Occipitotectal and Occipitocollicular tract

Frontotectal tract

Pretectal nuclei

Visceral nucleus III nerve

Superior colliculus

Oculomotor nucleus

III nerve

Trochlear nucleus

IV nerve

VI nerve

Abducens nucleus

Vestibular nuclei

Medial longitudinal fasciculus

Illustration of the multileveled interconnections, from the upper thoracic cord to pre-frontal cortex, serving pupillary, verging and binocular eye movements. While extensive in distribution, they are primarily subconscious (or preconscious) in operation but essential for conscious processing. [From Guyton pp. 575 & 761, with permission]

Many years of basic physiological research have established the sequential or near simultaneous activation of centers within the thalamic (hippocampal) frontal regions and upper spinal cord when the human eye(s) interacts with the environment.[9,10,11] The reactions upon which much research has been done include pupillary constriction/dilatation, lens accommodation, extraocular scanning and tracking patterns, and saccadic movements.

When the pupil is actively constricting or dilating in response to changing environmental illumination, we can be assured by a multitude of studies that this is indicative of multi-leveled, coordinated activation of the central nervous system in regions extending from the upper thoracic intermediolateral cell column (origin of the cell bodies of the sympathetic nervous system serving the eye via the superior cervical ganglia) through midbrain and the thalamic portions of the reticular activating system to the pre-frontal cortex. (See Figures 5a and 5b)

Similar widespread multi-level activation takes place (a) when the lens of the eye is actively accommodating to focus the sharpest image on the retina, (b) when the extraocular muscles coordinately function to converge or scan, and (c) when saccadic movements are taking place as in reading or scanning a broad horizon.

The activation of the reticular activating system (its generalized and more focal thalami components), which is a prerequisite for the appropriate initiation and continuance of each of these aforementioned functions, is a neurophysiological description of the process referred to as attention.[12]

Without this activation to a state of attention, higher centers are not brought beyond a minimal level of functional participation in processing visual data. Thinking, as we commonly understand this term, simply does not take place. (See Figure 6)

Extending throughout the midbrain and thalamus (in both its general and localized nuclei), the reticular activating system serves as the primary locus of attention. When it is not activated, higher centers are not engaged in functional conscious activity.

Seen in the context of human behavior, this becomes self-evident. The human being, since his appearance on the planet, has relied heavily on his visual system for survival/pleasure

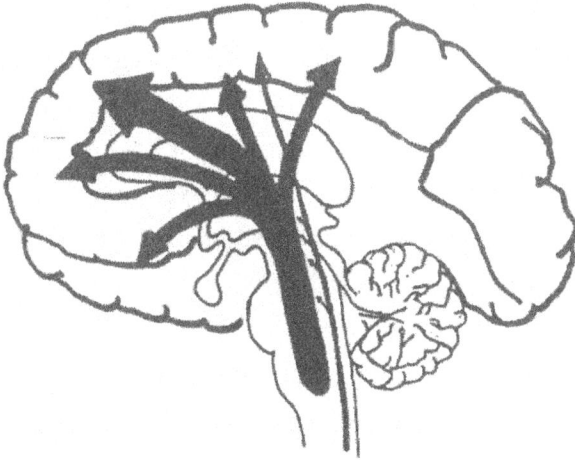

Figure 6
Illustrating the reticular activating system

(and all their derivative and associated human experiencings). These mechanisms, prewired and only marginally negotiable through experience or intention, operate at the leading edge of the human being's interaction with the world about him. The hunter-gatherer (male or female) has these multi-leveled and harmonized systems functioning at their optimal level when the environment is actively engaged. The mechanisms of pupillary dilatation-constriction, lens accommodation, extraocular move-ment, and saccadic movements are fundamental and essentially non-inhibitable, making it possible to search out, scan, focus, and identify whatever comes within the visual field. Without them early man would never have survived.

Parallel with the activation-response of these mechanisms is the near instantaneous and continuous activation-response of the reticular activating system, creating the attention which wakes up each of the many associated higher centers, making possible the widespread sharing/integration of visual information, which forms the substrate for yet higher integration leading to decision and bodily action.[13] (See Figure 2) By higher integration is included the functioning of the visual association area, the common integrative area, and ideomotor center (left cortex, areas 39 and anterior to it).[14] (See Figures 6 and 7)

Electroencephalographic studies have demonstrated this process of cortical activation in response to the aforementioned visual mechanisms to such a predictive degree that in clinical practice the presence or absence/modification of these visual mechanisms is taken as a direct indicator of the level and degree of cortical activity.[15]

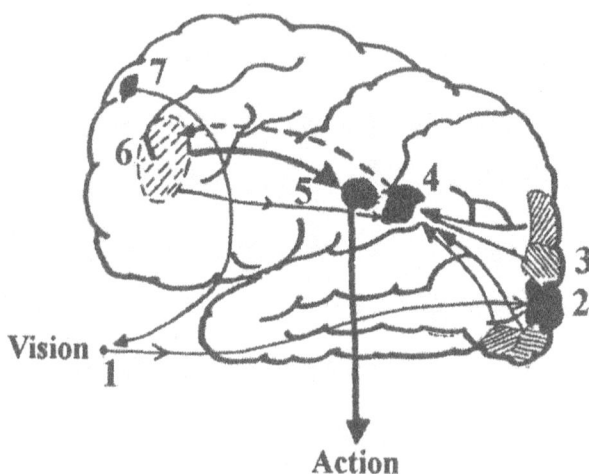

Figure 7
Visual Integration, illustrating the flow and integration of the process of visual perception.

1) Retina - Optic N - Optic Radiation to 1° Visual Cortex. (2) 1 Visual Cortex. (3) Processing through the associative visual cortex, the flow continues anterior to visual interpretive areas to (4), the common integrative area (area 39, left side of brain in 90% of humans). From (4) information is shared to and from the frontal cortex (6) to return to the ideo-motor center (5), the final common pathway on the road to "decision to act." (7) Prefrontal connections to the eye

We can also make use of the image of the hunter-gatherer to explore two corollary characterizations of the human being's external senses. Our external senses accept what they receive as the "truth." Put otherwise, all of our senses are innocent. They each have prewired constructs which enable them to interface with a particular form or energy coming from our environment. At this interface a transformation occurs which makes it possible for a resonant representation or energy analog of the external energy or form to be presented to the central nervous system.

The prewired state of each of the senses is a dependent, acceptant, "innocent" set of transformers so put together that

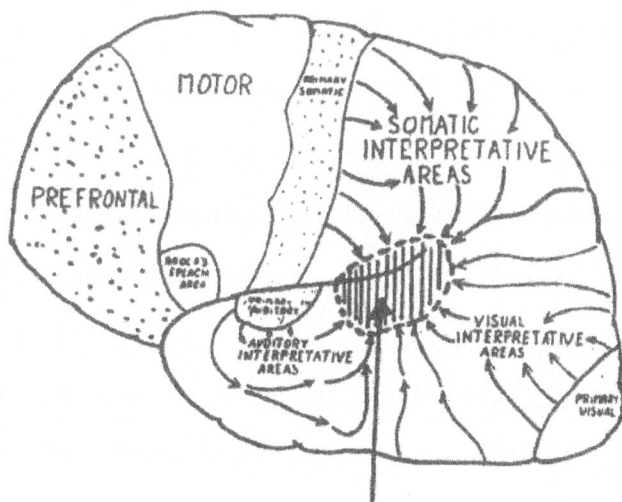

Ideo-Motor and Common Integrative

Figure 8
Illustrating the critical importance of the common integrative area (Brodmann's #39) and, immediately anterior to it, the ideo-motor center. All sensory data is confluent on these areas, making possible the construction of wholes from the various levels of input. These serve as the basis or the primary determinants of the course of action to be followed.

no option or selectivity is allowed within the system. It must respond to the energy or form presented to it wholly. What is real, for the central nervous system, is this energy analog or resonant representation. The form and sequence of the energies presented to it is accepted by the sense receptor as the truth. (In the case of the visual system, this is obviously a complex reflectant patterning of the visual portion of the electromagnetic spectrum.)

It is also important to note that the external senses accept, or take in, the spectrum of energies presented to it as a patterned whole.[16] In the case of the visual system at its point of interface (the retina), the complex, multiwoven, and changing pattern presented to it is a whole something.

Further along in the processing of the transformed visual neural analog, various portions of the whole are recognized[17] (e.g., angles and edges, color, motion), but it is a mandatory observation that the functions of the various associative centers and their integration have one primary goal. That goal is to construct a resonant representation (within the brain) of the electromagnetic energy pattern presented to the eye.[18] It is within the above context that we can state that each of the external senses (1) must believe what is presented to its interface and (2) that it perceives it as a whole.

Summarizing the sensory status of our hunter/gatherer, we can state that:

1. A number of visual mechanisms (pupillary responses to light, accommodation, extraocular and saccadic movements), when operating, are direct indicators of cortical, subcortical, and spinal activation, which extend across and through vast distances within the central nervous system.

2. Most of these activated areas are intimately involved

in the processes called attention and vigilance, which must be operant before higher cortical awareness and processing can take place.

Deeply embedded in these processes, and minimally modifiable by the function of higher centers, is the unqualifiable paying attention to change that appears in the visual (or other special sense) data. When something changes in the visual field (moves, becomes brighter, alters form, etc.), the eyes are "preprogrammed" at this deep level to follow or respond to that change. The hunter will pay attention to that change, not from conscious choice, but because paying attention to change in the visual field is the expression of an ancient and matured survival skill. As referred to above, this can be modified for brief periods through intentional over-riding, but these, for our discussion, are short-lived and rare events.

The external senses can be said to be innocent in the sense that they accept the incoming energy unqualifiedly. It is true in the same sense that a fiction told to a child is accepted as the truth. It is what it is; it is not a trick, a ruse, a partial presentation of reality. Our senses function analogously.

At the same time, the presented energy or form is accepted as a whole (e.g., a tree or an animal may be visually focused on by our hunter/gatherer, but the tree and the animal are embedded in a matrix of relationships: the color, size and position of rocks, other trees, the ground, hills, the sky). It is the whole of this that is presented to the eye in its electromagnetic analog of reflected energies, and it is only in the context of this whole that our hunter/gatherer will respond. More will be said in the next chapter about the richness of this reflected light environment and the long historical development of the visual system within this varied and subtle environment.

Reflected and Radiant Light Sources

The information in ambient light, along with sound, odor,
touches, and natural chemicals, is inexhaustible.
— J.J. Gibson[19]

In the preceding chapter reference was made to reflected light as the source of the electromagnetic array of energies presented to the visual system. Reflected (ambient) light is the source of all visual information that the human being and all other life forms with visual systems have responded to since the appearance of such forms hundreds of millions of years ago. This fact is of fundamental significance.

The subtle and multiple evolutionary changes that are so well documented in microbiological, plant, animal, and human studies speak with undeniable eloquence of the constant interplay/relationship between a respective life form and its total environment. Recent studies focusing on the beautifully adaptive changes in nuclear and mitochondrial DNA illustrate the creative and mirroring dance that can always be found between the whole of the determining conditions (the whole environment) and the whole of the particular system under study.[20,21,22] This is equally true of the visual systems of all life forms. From the simplest manifestation of light sensitive tissue to the complex mammalian visual system, each step has been tentatively and then more assuredly taken within a contextual whole of reflected light. For our purposes the differentiation between an external sense and a perceptual system as made by Gibson is now essential to note.

1. "A perceptual system is defined by an organ and its adjustments at a given level of functioning, subordinate or

superordinate. The organs of the visual system, for example, from lower to higher, are roughly as follows: First, the lens, pupil, chamber, and retina comprise an organ. Second, the eye with its muscles in the orbit comprise an organ that is both stabilized and mobile. Third, the two eyes in the head comprise a binocular organ. Fourth, the eyes in the mobile head that can turn comprise an organ for the pickup of ambient information. Fifth, the eyes in a head on a body constitute a superordinate organ for information pickup over paths of locomotion.

"Accommodation, intensity modification, and dark adaptation go with the first level. The movements of compensation, fixation, and scanning go with the second level. The movements of vergence and the pickup of disparity go with the third level. The movements of the head, and of the body as a whole, go with the fourth and fifth levels."

All of them serve the pickup of information.[23]

2. "A special (external) sense is defined by a bank of receptors or receptive units connected with a so-called projection center in the brain. The receptors can only receive stimuli passively, whereas in the case of a perceptual system the input-output loop can be supposed to obtain information, actively ... A (visual) system can orient, explore, investigate, adjust, optimize, resonate, extract, and come to an equilibrium, whereas a sense cannot."[24]

As mammalian evolution is traced, we can identify many developmental steps of the visual system up to its present form in man. At every point along this journey it is essential to hold in mind the physical presence of the life form within the encompassing atmosphere of ambient light.

Each new synaptic junction or associative center relationship that can be focused on within the whole visual system represents a movement to maximize the breadth of information gathering potential—always with the presumption of its relationship to the environment of ambient light.

The extraordinary richness and complexity of this ambient light environment requires for our purposes a bit of exploratory explanation. Figure 9 illustrates the origin and emergence of the ambient light environment. As radiant light from our sun enters the upper atmosphere, it is partially absorbed, reflected, or refracted as it penetrates toward the earth's surface.

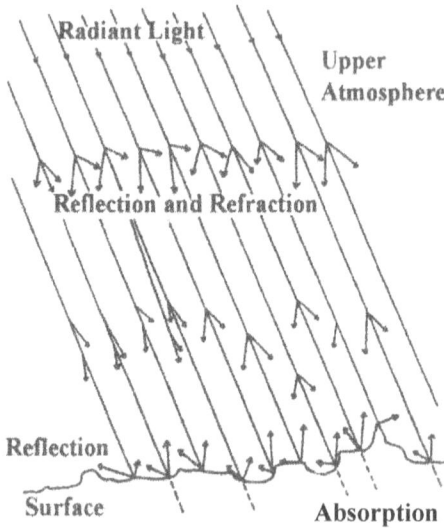

Figure 9
[After J.J. Gibson, with permission]

The varying textures and surface contours of the solids, liquids (and gases to a lesser degree) on and above the earth's surface create a mosaic of infinite potential, which forms the ground into which radiant and reflected light move. At the surface a symphony of infinite reflectance-absorption takes place. The resultant array fills the space in which life forms are present. Each point within this space is crisscrossed by an infinitely complex and steady state of reflected light. Wherever a life form moves, wherever it casts its gaze, there is this electro-magnetic array, constant and coming from all possible directions, that realizes the source of all visual information. As P. Brou et al. noted, "Color vision, as a useful faculty, evolved in a primeval world in which light from the sun—scattered, refracted and reflected—was the chief illuminant."[25]

It is not possible to say that visual information comes from somewhere. As Gibson notes, "The information in the sea of reflected energy is not conveyed. It is simply there."[26]

Radiant Light

We will briefly characterize certain features of radiant light.

A. It has a specific area (or point) of origin.

B. From that origin it moves outward in all available directions and along essentially straight lines. From the perspective of a life form on earth vis-à-vis the sun, we can say that all rays of radiant light are parallel.

C. All radiant light originates from atomic (electron-higher to lower shell-quantum release of photon) or nuclear (fission-fusion) processes.

D. Radiant light contains no information in the sense we have been using that term in reference to a visual system. There is no intrinsic pattern or array to a radiant light source that can be presented as a whole to a visual system. Put another way, we do not look at a radiant light source (e.g., the sun, a light bulb) for information. Our information derives from the interaction of this radiant light source with our external environment producing the reflected light array.

E. Radiant light (portions of the spectrum) has a number of profound physiological effects on various life forms: (1) Vitamin D synthesis in the skin;[27] (2) the breakdown of bilirubin in infants (1° "blue" light)?[28] (3) influence of total daylight exposure on a type of depression;[29] and (4) chronobiology, the influence of radiant light on pineal and other gland function, and biorhythms of various systems.[30,31]

In each of the above examples the life form, or a specific part of it, is acted upon by the radiant light source. It is not a question of information being made available—rather more like an order being given to activate a certain physiological-biochemical process. We will return to this particular difference between radiant and reflected light later and explore its potential importance.

In summary we can state that in the natural world:

Ambient light is infinite in the variety of its resultant array. Radiant light has a uniformity which cannot be termed in any sense an array. It is (in our world) both simple and finite with respect to the degree of variedness it contains.

2. Ambient light is infinite in the amount of information that can be extracted from it—there is always more that can be seen in a given visual array. Single-sourced radiant light contains no information.

Ambient light is infinite in the reflectant sources of its origin. Radiant light has a specific source that is relatively invariant.

Ambient light is a source of potential information. Insofar as has been discovered, it does not function as an effector vis-à-vis biological systems. Radiant light is both the fundamental source of reflected light and, in its own character, it is an effector or activating influence in many biological processes.

With the establishment of the above mentioned parameters of difference between radiant and reflected light, we are brought, pragmatically, to the question of the significance of these differences. (I am speaking about both physical characteristics and physiologic effects of light which are a field of intense research at present. Further points of differentiation between radiant and reflected light will doubtless emerge from this fertile activity.) Stated as questions we may ask, "What difference does it make to the visual-perceptual system and the central nervous system of a human being?" "Are there demonstrated neurophysiological differences, and are these differences of any significance to human health?" One of the fundamental premises of this book is that such neurophysiological differences do indeed exist, that to a definitive degree they have been demonstrated by research, and that their implications vis-à-vis human health ("hale," that is, whole in Bohm's terms) are considerable in number and pathological (dysfunctional) in their potential and real influences.

Perceptual-Cortical Disharmony (Dysfunction)

Perceptual variety is basic to the motivational and emotional states of the individual. – W.N. Dember

The greater the interrelations in terms of synaptic contact the richer, the more intelligent the mental life.
                                     – F.E. and M. Emery

Choice of Futures
The review and exploration of the pertinent literature will occupy the largest part of the present chapter. At its outset we will state a number of conclusions drawn from this literature and then proceed to document the reasons leading to these conclusions. For much of the data, charts, and literature citations, we are indebted to Merrelyn Emery, PhD, of the Australian National University at Melbourne.[32] Dr. Emery has gathered, and

interpreted in great detail and with telling effect, the literature that deals with cathode ray technology, radiant and reflected light, and television as a marketing, entertainment, and educational tool. Her approach, from the perspective of Sommerhoff's model of adaptation, is a systems analysis demonstrating great power and perception.

Television viewing, when taken as a form of perceptual-cortical interaction with a radiant, repetitive, simple signal source, results in multi-leveled perceptual-cortical disharmony (maladaptive in Emery's terms).[33] This general statement derives from the following closely related and somewhat overlapping observations:

A. Established indicators of pre-cortical and cortical activation are effectively nullified by television viewing. This includes previously mentioned alerting and vigilance mechanisms tied to pupillary dilatation responses, vergence mechanisms, and extraocular tracking and saccadic movements. These mechanisms operate in the pre-conscious processing time of Libet[34] and qualify all further processing.

B. Marked contraction of the full potential visual field of binocular vision takes place. This reduces the total visual area of registration (context) and fuses the effective matrix or ground (the equivalent of the total peripheral visual field) and the supposed content of the visual image itself, i.e., the image focused on is embedded in a matrix of repetitious, simple, radiant light signals. This represents, therefore, a contraction of the source of visual information to a very small portion of what the human visual system is both capable of and for which it has been prepared to deal with through evolution. Closely linked to this form of deprivation is that derived from the physical nature of the color signal.

C. As will be discussed later in more detail, the color of a television image is the result of phosphor activation by a stream of electrons. The wavelengths of visible light deriving from this physical interaction are (a) only a portion of the potential in the visible spectrum, and (b) combined in what can be termed extremely simple combinations (when compared with the reflected light array of the natural world) to produce the different colors we see on the screen. Both of these features contribute to a paucity of stimuli when measured against the natural world.

D. The majority of the sub-organs (as described by Gibson)[35] which comprise the visual system of the human being are rendered inoperative, i.e., pupil-lens-retina, binocular vision, eyes in a movable head, and head on a movable body. This further deprives the visual system of a diversity of information required for real responses to the real world. Additionally, there is the reduction in input from the physical body (via external movement and proprioception), which contributes to the flow of input into the reticular activating system, thence further affecting attention and vigilance.

E. There is an absence of "satiety" recognition in television viewing. In normal human activities (e.g., eating and sex) there is a built-in satiety point which turns off the activity. For a variety of reasons to be discussed, this appears to be lacking as a physiological endpoint with respect to television viewing. In this sense, as Pawley[36] notes, "Television is endless." This feature appears closely related to Emery's contention that "Television viewing is goal-seeking but purposeless. Its end is in its immediate consumption."[37] The reward is in the viewing; its motivation is to continue viewing.

F. During viewing itself an intentional reduction or elimination of other sounds, of physical movement (in fact the frequent assumption of a maximum ease posture) greatly reduces proprioceptive input. The remarkably poor quality of sound in most television sets also reduces to a small fraction the input potential that the human ear can process.

G. From the limited processing of a television signal that does occur, output from the reticular activating system descends through the spinal cord, and output from nuclear origins of the sympathetic nervous system moves into the body matrix via neural and vascular (hormonal) channels—both in order to prepare the body for action. But appropriate action-response does not occur while viewing. A cycle of stimulation-preparation-frustration ensues at this autonomic level, perhaps only to be released when the television is turned off.

H. Color television adds a further dimension to the disharmony. There are pathways for higher color processing (also for the processing of music, tone of voice, spatial relations and especially, the recognition effaces) in the right hemisphere.[38] This processing does not require the predominantly left hemispheric functions of logical analysis, comparison, verbal recognition, etc. In fact we will present EEG evidence that indicates clearly the singular reduction in left hemispheric functioning that results in the right hemisphere functioning in quasi-isolation. These two facts enhance the "feeling tone" of familiarity without real knowledge that is characteristically demonstrated when testing-by-recognition is contrasted to testing-by-recall.

I. Evidence points to the distinct possibility that, as higher cortical function is impeded, subcortical, and in particular, hippocampal (limbic brain) functioning may be more in evidence. "While the neocortex rests or sleeps, the old brain comes out to play."[39] Hippocampal and associated areas "are implicated in the dreamy state, a kind of double consciousness … where the individual has the sense of being in contact with reality, but at the same time has the feeling he was experiencing a dream or something that had happened before."[40] This dissociation or disharmony between the old brain and the higher cortex has implications extending to effects on normal dreaming as well as on the perception of reality through this dreamy state.

Central Nervous System Responses to Television Viewing

There is within the organization of the human brain, a close and fundamental relation between vision, verbalization, consciousness and the purposeful nature of man. – Emery

Extraction of meaning is directly related to cortical arousal. – Emery

With these summary conclusions in mind we will proceed to a review of the pertinent literature. To the extent that studies have been carried out (the relative paucity of the research literature has been referred to previously), the primary evaluative tool has been the electro-encephalograph (EEG). Scalp leads (isolated and multiple) have been utilized exclusively, and, unfortunately, there has been a lack of uniformity in procedure, in the number of leads simultaneously recorded, in the distribution of subjects as to age-sex-health status, in the length of time recording was carried out, and in comparator EEGs (reading in ambient light vs. backlit screens). In spite of these non-uniform factors, the findings themselves are provocative enough to warrant careful study. As frequent reference will be made to a variety of EEG findings, it may be useful to mention certain electroencephalographic "benchmarks."

A. "The ascending reticular activating system is an important source of signals that excite the dendritic layer of the cortex. A close relationship therefore exists between brainwave activity and activity in the brainstem or thalamic reticular activating system Scalp electrodes will reflect primarily the state of activity in the dendritic layer of the underlying facilitated cortex and its associated portions of the reticular activating system."[41]

B. The faster the brainwaves, the more intense the level of activity in the underlying cortex. With mental activity the brainwaves also become, generally, more synchronous.

C. The slower and more synchronous the brainwaves, the less active is the cortex being monitored.

D. Functionally specific areas within the brain are fairly well known (e.g., 1° visual cortex and the associative visual centers located in the occipital lobes; verbal associative areas in Broka's convolution). EEGs taken with electrodes over these areas are quite indicative of the degree of activity within the cortex being monitored.

E. Brainwaves are categorized into four main divisions:

Delta: All waves below 3.5 cycles/second. Found in deep sleep, organic damage, and when thalamocortical connections are severed.

Theta: 4–7 cycles/second. Found in children, during emotional stress in adults, and in many brain disorders. Found in resting, awake state relaxed. Disappears in sleep.

Alpha: 8–13 cycles/second. Found in resting, awake state relaxed. Disappears in sleep.

Beta: two types—(a) *Beta I:* Frequency approximately twice Alpha. Disappears with mental activity; and (b) *Beta II:* Frequencies as high as 25–56 cycles/second. Appear with activation of the central nervous system; are synchronous and of low voltage.[42]

In addition to EEG studies, we will refer to studies done on regional cerebral blood flow ($_rCBF$) which measure available O_2 (O_2a). While studies taken during television viewing have not been done using $_rCBF$, a number of studies contrasting radiant and reflected light sources have been done, and these will serve to emphasize the brain's quite different responses to these sources.[43]

At the end of the book, reference will be made to newer investigation techniques which hold much promise in clarifying the neurophysiological processes involved. Commercial television emerged in the immediate post-World War II years. Slowly at first and then with rapid acceleration, it has continued its meteoric rise to its present status as the premier source of news and entertainment for the American public. Nielsen reports that in 1997,[45] 97 million United States households (262 million persons as viewers) had television sets; 96% of these had color television, and 57% had two or more sets.

In spite of the meteoric rise in use from 1950, it was 1971 before the first investigator recorded an EEG on a person viewing television.[46] When Herbert Krugman studied this first EEG, he was so impressed by the findings that he published a paper on this single subject. Because it was the first such paper, we will highlight the process of his thinking.

Krugman was led to the use of the EEG from prior studies which noted a marked difference in "personal life connections" made by subjects when reading print and when viewing similar advertising on television.[47,48] In these prior studies, it had been found that subjects made markedly fewer personal life connections (associative thoughts with a personal memory component) when viewing a television advertisement than when reading the same advertisement in print. It was this difference that interested Krugman and led him to review the literature that made use of the Mackworth Optiscan[49,50] to compare the degree to which subjects differed when scanning (the application of saccadic movements) an advertisement.

Conclusions from these studies indicated to Krugman that "scanning, as an active learning process, had something to do with the possibility of personal connections taking place between the stimulus and viewer.[51] This, Krugman felt, represented a demonstration of at least two types of attention—one involving considerable effort on the part of the viewer and the other involving little or no effort. This led him to his question: "What is there about the changing stimulus in television that can relieve man of the work of learning? Is learning the right word to use, even qualified as passive learning?"[52] From these questions and other observations which noted a singular lack of pupillary response to television ads, Krugman proceeded to raise the issue of possible brainwave changes that could help to explain this passive learning.

His single subject, a 22-year-old female secretary, was studied using a single occipital electrode. Multiple viewings (3 each) of three commercials were used as test material. Several conclusions were drawn:

1. The response to print (ambient light on a printed page) "can be considered active"[53] and primarily composed of fast waves (12.3 to 31 Beta c.p.s.). The fast waves were consistently five times the amplitude of television-evoked waves (slow, 1.5 to 7.5 c.p.s. and Alpha, 7.16 to 12.33 c.p.s.).

2. In terms of percent of time occupied by the three EEG wave bands, the following is a graphic display:

	Reading Hard Copy Reflected light	Watching TV
Slow Wave - Delta (4-7 c.p.s.)	10	46
Alpha(8-13c.p.s.)	32	30
Fast Wave - Beta (13+c.p.s.)	56	24

3. Within 30 seconds (Krugman calls this the characteristic mode of response), slow wave activity during television viewing came to predominate.

4. With each repetition the slow waves gradually increased.

Krugman himself summarized these findings by stating that "the response to television is more passive simply because it is an easier form of communication." The term "easier" here refers to his earlier division of attention, this being representative of that attention requiring little or no effort.

What is of primary significance in Krugman's study is his demonstration of the dramatic difference in the degree of central nervous system activation between reading print and watching television. He is the first investigator to note that viewing television produces what on the surface appears to be a paradoxical response, namely, that the occipital cortex is markedly reduced in its level of activity when all prior EEG investigations had highlighted the marked increase in occipital cortical activity when learning by visual means takes place.

Keeping in mind that a study of one subject using one electrode does not form a firm foundation for conclusions, we will review a number of studies that followed Krugman's publication.

Rossiter had previously (unknown to Krugman) tested television (radiant light) and film (reflected light on a screen) using pupillary dilatation responses as the primary comparator. He found "significantly lower pupillary dilatation responses to the same film presented on a television screen than when presented on a film screen even though image size and light intensity were identical."[54,55] Studying pupillary responses as an index of cortical activation and vigilance validated the conclusion that television viewing produced a significantly smaller pupillary response (and hence a smaller or less active cortical response) when compared to viewing the same material on film.

We referred earlier to studies of saccadic movements as quantitative reflections of central nervous system activation, namely, that a higher number of saccades would indicate more intense cortical activation, fewer or no saccade indicating the opposite.[56] The final technical report found a ratio of 5.7–9.2 to 1 in the number of saccades while reading compared to television viewing. Featherman also noted "a significant decrease in both theta and beta activity during the television versus the reading condition, suggesting a reduction in cortical arousal during television viewing," and "a significant higher amount of theta (slow) activity was observed in the left hemisphere across all conditions."

Concerning the decreased activity levels noted over the left hemisphere during television viewing Krugman states: "A gross total record of brain activity over time showed that the left hemisphere tires and gives way to the right, eventually to the point of achieving 'natural' viewing, with the left hemisphere 'turned off' and the right hemisphere remaining alert."[57]

In 1979 Appell et al. reanalyzed data from their EEG study of thirty right-handed women at the request of Krugman. They found that, in each of their three trials of EEG recording while watching television commercials, "left [hemisphere] dominance declined exponentially over time."[58]

Silberstein et al. undertook to test brainwave responses to television viewing compared to other reading/viewing activities. Unfortunately, from their description of test conditions, it would appear that all four conditions were variations of radiant light and hence not an ambient-radiant light comparison. In spite of this they note in their summary for all four conditions: "For neither hemisphere in relation to any of the contents was there a sign that television viewing, or viewing radiant light, had the power to spark sufficient beta activity to indicate predominantly intellectual activity."[59]

Correlating visual mechanisms previously discussed as primary indicators of cortical arousal with EEG findings, Mulholland notes that "the best way to predict if alpha (slow) will be more or less is to evaluate the degree of visual control, i.e., the number of reflex systems (pupillary dilatation, lens accommodation, converging or tracking movements) involved in maintaining the best vision."[60] In his study he noticed that, "children watching television often drop to a rather low level of arousal, with plenty of alpha. The posture is often a relaxed one, especially the facial musculature. This high level of alpha led me to speculate that children may be spending a huge amount of time *learning how to be inattentive*."[61] (Emphasis mine.)

In corroboration we would cite Erik Peper's comment to Jerry Mander.[62] "Any orienting outward to the world increases your brainwave frequencies and blocks (halts) alpha wave activity. Alpha occurs when you don't orient to." Later in the same statement Dr. Peper says, "Instead of training active attention, television seems to suppress it."

We are now in a position to bring together and extend certain of the observations made in the segments of this chapter dealing with radiant/reflected light and central nervous system responses to television viewing.

Figure 10 is a graphic compilation (after Emery)[63] of 47 studies which involve a variety of learning behaviors under reflected/radiant light conditions. EEG monitoring clearly demonstrates a marked preponderance of slow wave activity (lower levels of central nervous system activation) under radiant light conditions. Behaviors included mental arithmetic, writing a complicated word test, televised word text, mental imaging, acoustical tracking, verbal listening, reading hard copy, eyes open, visual reasoning, and mentally composing a letter.

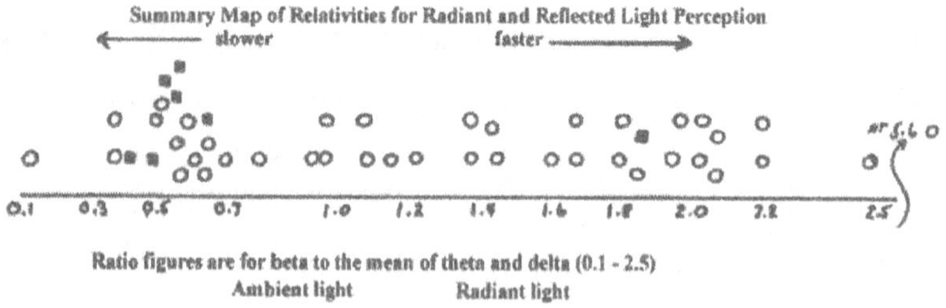

Figure 10
Comparison of 47 brain wave studies using radiant or reflected (ambient) light sources

Van Lith et al.[64] compared the evocation of cortical potentials by projector and by television. They found that latency (delay) times with television were almost twice that with projector. They concluded "that the 50 Hz signals, probably via the television set, seriously impaired the evoked potentials."[65] Clearly this confirms the fact that the central nervous system of a human being responds differently to radiant (especially pulsed) and reflected light, and that this difference is at a neural (subconscious) level. As noted by R. Ackoff and F. Emery in 1972, "The nature of the physical information contained within the radiant light signal will have direct effects, not at the level of learning or meaning, but at the level of neuromechanism or CNS reaction rather than purposeful response."[66]

In summary, there is evidence from EEGs and other studies taken during television viewing that:

1. Visual attention mechanisms are severely dampened.

2. The left cortex (hemisphere) decreases rapidly in its level of fast wave activity to the point that within a short period (30 seconds by Krugman), slow wave activity greatly predominates and the left hemisphere is effectively "turned off."[67]

3. Over all cortical areas there is no evidence that predominantly intellectual activity is taking place.

4. The right cortex, continuing to process the television image, but with evidence of some decrease/alteration in EEG fast waves, does not cross-relate its activity with the left cortex.

Regional Cerebral Blood Flow

The concept of metabolic activity in the brain being reflected in available oxygen (O_2a as measured by regional cerebral blood flow) has led to a clearer anatomical-physiological understanding of the function of specific areas within the brain. It is more precise, anatomically, than EEG studies and gives a real-time reflection of changes in focal area function than were previously impossible to obtain without much conjecture. (Newer investigative techniques such as expanded electroencephalographic recording, PET [Position Emission Tomography] scanning and dynamic MRI [Magnetic Resonance Imaging] studies have the potential to clarify the process of cortical and subcortical dysfunctions to a far more specific degree.) Most of the studies utilizing this technique are directed at elucidating known pathological states (e.g., Parkinson's disease, schizophrenia) but a few are pertinent to our discussion. Cooper et al.[68] compared stroboscopic illumination (radiant light) with a presentation of pictures viewed in reflected light. They found a rapid rise (up to 20% of available O_2 in area 19 (associative visual area) under reflected light conditions. No change was noted under stroboscopic illumination. Simultaneous EEG recording noted much more de-sychronization of alpha during picture viewing than during radiant illumination.

Further, Cooper compared patterns of increasingly natural content and came up with these findings:

Meaningfulness (by O_2a levels)

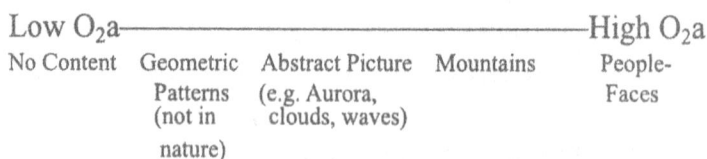

Low O_2a				High O_2a
No Content	Geometric Patterns (not in nature)	Abstract Picture (e.g. Aurora, clouds, waves)	Mountains	People-Faces

Figure 11
[after Emery] [69]

In a similar study Phelps[70] reported that radiant light produced small increases in metabolic activity in the primary and associative visual cortex. An increase in pattern (with radiant light) increased the activity in the associative cortex—an expected finding given our understanding that the associative cortex is concerned with more complex pattern interpretation. Cooper found a tenfold increase in activity of the associative visual cortex under natural viewing conditions (reflected light) of a park. These observations are strongly supportive of our previously emphasized view of a visual system which has evolved to take in and give meaning to wholes. In point of fact, the more complex the whole, the higher the level of metabolic activity observed in the associative visual cortex.

In further confirmation of these points we would direct the reader to the published work of J. Engel[71] and Simon.[72] Simon's work in particular is cogent to our theme as she clearly demonstrated, via EEG continuous recordings over a 48-hour period, that the human being spends the majority of his waking hours in an alert, active state of environmental scanning (vigilance), a state which watching television ... appears not the induce."

Habituation - Radiant Light - Repetition

Habituation of response to a stimulus diminished conscious attention. – Emery

Merrelyn Emery, in her thesis,[73] appropriately raises the question as to whether true physiological habituation may not be taking place during television viewing. Anecdotal[74] and peripheral research commentaries[75] have raised the same question for many years, but Emery has brought together a formidable array of research reports to support this habituation hypothesis. As she notes, "Habituation is the process whereby the desynchronization of cortical EEG produced by novel stimulus tends to shorten and gradually disappear with repeated non-reinforced stimulus presentations." With respect to visual systems, there is a large body of photic stimulation research which establishes certain characteristics of habituation to a light signal. Summarily, these state that the more simple (less complex) and rhythmic the stimulus, the more rapid and complete the habituation.

A television signal is, in the physical (electronic-photic) sense, a radiant light source producing a light stimulus in a steady repetitious cycle of 50 times per second (60 c.p.s. in some countries). Van Lith et al. recorded evoked potentials over the visual cortex occurring every 20ms (50/second) during television viewing. As he notes, "The picture on a television screen is built up by a light point which runs over the screen in 20ms. When the light point is outside the evoked potential visual field, the screen is dark for the electrode which records the evoked potential. When the light point runs inside the E.P. visual field, the screen lights up in relation to the electrode placed over the visual cortex.

This implies that every 20ms (50/second), the electrode over the cortex "sees" a light stimulus and registers an evoked potential, at least if the 50 Hz flicker is below the critical fusion frequency (CFF) of the visually evoked cortical potentials."[76] From prior studies it has been presumed that the CFF is around 50 Hz in humans. Emery quotes research reported in the *New Scientist*[77] however, that has shown "that single nerve cells from the optical tract of a cat can 'lock in' to the florescent flicker and pass signals on to the brain at the 100 Hz frequency." Verification of this physiological possibility in the human being needs to be done, as it would greatly clarify the issue of television-induced epilepsy as well as strengthen the habituation hypothesis of Emery.

Habituation via the television signal offers a clear and comprehensive understanding of the findings previously noted. The visual signal, passing into the optic nerve, will have its first level of synaptic relationship with the lateral geniculate body and from here be shared into the reticular activating system with a number of midbrain, spinal cord, and subcortical centers.

The repetitious, rhythmic, and simple impulse would rapidly desensitize these centers (they would quickly conclude that nothing important is going on, and so the attention/vigilance mechanisms would be dampened).

Emery cites evidence that points also to higher pre-frontal and associative area domination by the signal, which effectively would lessen or block their normal contribution to processing prior to involvement of the common integrative and ideo-motor areas, from which "decision to act" emerges. (See Figure 9) Without appropriate pre-frontal processing these "final common pathways" to decision and action are flooded by sensory data and would tend to initiate impulsive or unjudged responses to this sensory data. While these two levels of habituation were operant, the image, more powerful yet if in color, would be presented to the right hemisphere and hippocampal structures.

With the suspension of normal attention/vigilance mechanisms, the left hemisphere would minimally participate in evaluating the full content of the image. The affective/emotionally evocative potential of the hippocampus and right hemisphere would be unrestrained by the logical-analytical, verbal capacity of the left hemisphere. The "dreamy state" of MacLean[78] could be the result. The cited literature detailing visual response mechanism changes, in EEG, and evoked potential observations while viewing, and the differentiation of the natures of radiant and reflected light are all consistent with this habituation hypothesis.

Television Epilepsy
The Far End of the Pendulum

The occurrences of seizures induced by viewing television were first reported in 1952. Since then a considerable body of research literature has accumulated which has confirmed the reality of this disorder and defined a number of its fundamental parameters.[79,80,81,82,83]

There is still disagreement, however, as to the relative significance of flicker, pattern, and/or kindling as induction factors in this disorder. From the literature cited here, it would appear that all three may be involved to a greater or lesser degree, in differently sensitive individuals. As Emery notes, "The range of cases illustrated that any one or a combination of radiant light source, flicker frequency, or certain patterned stimuli can be responsible. It becomes immediately apparent that television is often, and most usually, an excellent, integrated source of all three ... a highly provocative stimulus."[84]

For our purposes it is not essential to review the literature in detail. It is sufficient to state and corroborate the reality of television seizures and to summarize a number of factors that have been recognized as a result of twenty-five years of research. It is of more than passing interest that in 1959 one author (Klapetek)[85] expressed his concern about the rapid growth of television in view of its dramatic central nervous system effects— with obviously little effect on, or response from, the scientific or medical communities.

In summary, the findings in television-induced epilepsy are:

Seizure activity has been noted to occur: in sensitive individuals, while viewing a television screen, and during video games. The seizure phenomenon is, then, related to cathode ray technology, not only to television viewing.

Beginning from a broad definition of photo-sensitivity, Wilkins placed at risk 10% of adults and 15–33% of children. This is based not on convulsive seizures alone, but on a variety of unusual EEG responses to pulsed light. It may well be found that many of the non-seizure responses are directly indicative of degrees of disharmony arising from the attempt of the CNS to process an essentially indigestible, or deficient, form of impression.

Jeavons and Harding report that 56% of patients had seizures only while viewing television. Of particular interest is their finding that, of this 50%, more than half had normal EEGs even with intense light stimulation.

The incidence of television epilepsy, extrapolating from several studies into the general population, is slightly greater than 1 in 10,000. It has been documented primarily in younger people (peak incidence 8–14 years), but cases have been reported occurring from age 2 to age 52. With a viewing population between 8 and 14 years of age of approximately 47 million in the United States, there are approximately 4300 youngsters who suffer from television epilepsy (*grand mal* seizure type).

Of potentially greater concern is the unknown incidence of *petit mal* or absence seizures resulting from television viewing. No large scale study of this form of television-related seizure could be found in our literature search.[86,87] The fact that children in this age group watch approximately twenty-five hours a week of television and that *petit mal* seizures would be difficult to observe in the most common viewing conditions results in a question that should be of medical concern.

It is known that sleep deprivation, certain times in circadian rhythms, and hormonal fluctuations can increase photosensitivity (measured by intermittent photic stimulation [IPS]).[88,89] These factors add a further degree of subtle interaction that can affect the age group most at risk.

"There are no sources of environmental visual stimulation other than the television that possess the eleptogenic properties of 50 Hz flicker and 25 Hz pattern oscillation."[90]

Adaptation to flicker does not occur.[91] While persons in countries which have television sets with a 60 Hz flicker complain about the flicker which they note when visiting 50 Hz countries, there is no evidence that people in 50 Hz countries have adapted. The same subliminal neurological consequences would still be operant.

9. Self induction of "absence" (*petit mal*) attacks during television viewing is a well documented, although rather rare (so far), phenomenon.[92] That this is reported by subjects as a "pleasant or nice" feeling[93] raises the question of how frequently the attacks may not be reported in the most vulnerable age group. Also of concern is the laboratory observation that *petit mal* is the most commonly observed form during IPS. Consequently, there may well be a submerged, higher incidence of television epilepsy of the "absence" (*petit mal*) type than of the documented *grand mal* type.

Chapter III

Television Viewing: A Form of Sensory Deprivation/Disharmony

A television image is never whole. – Mander

What we have reduced this world to! – Kirshnamurti

THE MOST FUNDAMENTAL of osteopathic medicine's precepts concerns the effort to treat the patient as a "whole." Within that holistic viewing of man, other primary precepts find their meaning,[94,95,96] i.e., the neuro-musculoskeletal system as the final "instrument" of the human being in manifestation; the essential role of the vasculature in health and disease; the balance or harmony required between all systems and parts in order to maintain health; the enormously subtle and complex powers of the organism to heal itself; and, largely still unrecognized, the role of "somatic dysfunction" as a reflector and initiator of internal disharmony or illness.

Any factor which inherently introduces an unbalancing, or disharmonious, influence to the body would be viewed, in the context of holistic principles, as a promoter of illness. Television viewing from our perspective is such a promoter.

In this segment we will attempt to outline and summarize the argument for concluding that television viewing is a form of severe sensory deprivation/disharmony, an activity that presents such a bleak and fractured landscape, and in such a contrived technical manner, that our sensory systems, associative and integrative cortical areas are, at various levels, starved, numbed, habituated, and prohibited from rebuilding a perceptual "whole." Consequently, they are prohibited from interacting with our world in any way that reflects the fullness of real meaning.

The Scene

Typically a person viewing television will (a) sit in a partially or wholly darkened room, (b) choose the most comfortable seating available, (c) reduce sound, other than from the television, as much as possible, (d) participate in little or no tactile, taste, or smell events (except during breaks).

By doing the above, a number of neurosensory changes take place. Peripheral vision is reduced enormously (many viewers eliminate it altogether by having no other light source in the room). When the television is turned on, the matrix, or background, of the primary image on the screen becomes the whole periphery. It will be processed, or rather an attempt will be made to process it, by the visual system as such. Also, with the television turned on, we can illustrate the marked reduction in the visual field via these simple diagrams (Figures 12 and 13).

Even a book, if taken as a comparable static visual field, presents a much larger field. Static here refers to the assumption that when reading a book, we keep focused on the page. This is not true, however. Many studies[97] have shown that, while reading, the eyes continue to scan well beyond the page and in all directions. Reference was made earlier to the great difference (5.7–9.2 to 1) in saccade between reading and watching television.

The emphasis on reducing other sound sources (recall the frequent and strenuous verbal objection when another viewer

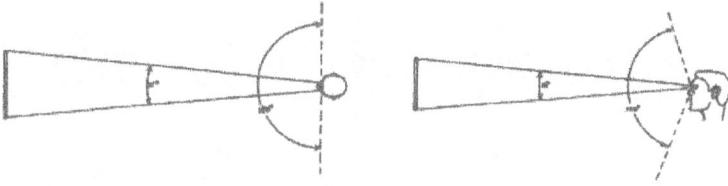

Figure 12

Illustrating the marked reduction in visual field utilized when viewing television. Typical viewing distance of 6ft– 7ft with 12ftx 16ft screen.

Figure 13

Illustrating the larger utilization of the potential visual field when reading a book. Note in the text the added difference in saccadic movements

or a child speaks up during the program) leaves the television sound as the primary source of auditory stimulation. In most television sets the quality of sound is very poor (this is obviously less true of some of the newer stereo systems). For technical reasons concerned with the proximity of the sound and visual signals as they enter a television set, it has been elected by the television industry to maximize the clarity of the visual signal. The sound signal suffers thereby. Compared to any ordinary whole human event (e.g., talking with a child, walking in the woods or on a downtown street), the sound coming from a television set occupies a very small portion of the potential auditory processing capacity. This applies both to the range of sound from the television and to the lack of other sound sources. Compared to real life events, it is not an exaggeration to term television sound deficient, or deprived, in comparison to the rich, natural auditory environment in which the human sense-brain normally delights.

The reduction in physical movement during television viewing also represents a considerable contrast to the great majority of other human activities. When we read, participate in conversation, or write a letter, we move a great deal more than when we watch television.[98] During reading, talking, and writing, proprioceptive feedback from throughout the body continually "feeds" the reticular activating system and higher centers, contributing to the state of attention and vigilance necessary for "whole" participation in an event.

So much for the scene.

The Switch

When a color television set is turned on, we, the viewers, perceive a "colored" image. This color is the result of high velocity electronic bombardment, in a fixed, predetermined fashion, with resultant excitation of three phosphors. These phosphors, present in a fixed dot-and-line pattern on the inside of the television screen, will radiate light of varying wavelengths when stimulated. (The phosphor information noted here was kindly provided by Steve Rand, a Phosphor Engineer at RCA). In most United States television sets, the red phosphor (yttrium oxysulfide) radiates red light in two narrow bands (615 and 625nm with a 3–4nm spread). The blue and green phosphors radiate over much wider bands.

–Blue Phosphor: zinc cadmium sulfide with activators, radiates from 400 to 550nm with a peak of 455nm.

–Green Phosphor: zinc silicate with activators, radiates 75% of its color from 510 to 590nm with a peak at 550nm.

–Blue and Green Phosphors radiate as a result of electrons moving between atoms, hence their broader band.

–Red Phosphor radiates as a result of single atom-electron shell activation and return to the rest state (quantum), hence its predictable narrow band.

By combining the activation of these three phosphors in various ways, a variety of "colors" is perceived by us (registered by the three varieties of cones). However, the simple experiment of looking back and forth from a television screen to any natural environment convincingly demonstrates that they are not at all the same. Large segments of the electromagnetic spectrum that the human eye is responsive to are not reproducible by the triphosphor technique of color television. Many ordinary and all subtle nuances of color and shadow are impossible to achieve. The author has questioned a number of people regarding this perception and the near uniform characterization of television color as "sharp," "harsh," "hard," "simple" when performing the subjective experiment of looking at, and then away from, a color television screen.

Again, in comparative terms, the variety of "color" (the breadth of the potential response of the human eye to the visual portion of the electromagnetic spectrum) in the natural world, the world in which, for millions of years, the human being's visual system has been adaptive, is immeasurably more complex and richer than what is presented by a color television screen. The color of a television screen is seen, as was the breadth of sound and the visual field dimension, as presenting a deprived and impoverished source of visual stimulation.

In this segment we must also highlight one of the differences previously noted between radiant and reflected light. Ambient light, in our natural environment, is a total, all-surrounding medium in which we live. It is present everywhere, and its rays enter the eye from all directions and at all possible angles. The human eye evolved, from far back in its mammalian past, within this total environment. It is co-adaptive, in every respect, with this environment. Ambient light is quite analogous to the air around us, or to the sea around a fish.

By contrast the radiant light projected from a television set is more like a focused jet of air or a stream of water from a garden hose. There is more here than simple analogy. The jet of air and the garden hose stream are both effectors; they bring about changes in what they impact. Mention was made earlier of the well-documented biological effects of radiant light on the human body.

One of the fundamental precepts of this book is that the human being's visual-cortical-endocrine system is affected by the radiant light from a television screen. These effects are suggested by a number of findings cited earlier, in particular those that point to attention/vigilance mechanisms that are adversely affected, and to the habituation of midbrain and subcortical centers.

Mention must also be made of the television image itself. Being constructed of a pattern of dots, it has limited power of resolution. Walk up to a television set and look at it from six inches distance (better yet through a low power magnifying glass). The clarity of the image is largely dependent on being viewed from a distance, the image itself quickly coming apart on close observation. The essential simplicity of television image construction illustrates the paucity of visual range given by a television image. Add to this the fact that a television image is never whole. While man may consciously see a whole image on the screen (due to 50 Hz flicker and 25 Hz pattern), there is evidence, cited earlier, that the image is anything but whole as perceived by lower centers and the visual apparatus itself.

In the lack of breadth of its photonic spectrum, its lack of a true three-dimensional diversity of origin and direction, a color television screen presents a flat-focused dimension of impoverished stimulation. It could be characterized as a junk food feast leading to visual indigestion.

The Process

As pulsed, radiant light from a television screen enters the eye, a sequence of disharmonious effects result. We must assume that, by prewired biological necessity, the central nervous system will attempt to digest or process this visual data. Its inability to do this appropriately (or "holistically") is demonstrated by the following effects:

1. Habituation of midbrain and thalamic portions of the reticular activating system, reducing attention vigilance mechanisms beyond their previous reduction by lack of visual and proprioceptive mechanisms

2. Marked reduction in left hemisphere activation levels. Inhibition of logical, analytical and verbal participation. "With only right brain processing operating, there is no mature purposeful course of action open to the individual."[99]

3. Some reduction in right hemisphere and hippocampal activity. A passive reception of the flood of images, with reduced processing. Possible habituation of the right hemisphere.

4. Lack of sharing across the corpus callosum and other commissure. A near "split brain" condition. (See Figure 14)

5. Excitement of unconscious, midbrain, sympathetic nervous system responses primarily to the superficial processing of images with all their emotional affect. A stimulation-preparation, frustration-to-action cycle results.

6. Marked reduction in overall body activity, both a result of 1, 2, and 3 and a source of reduced feedback of proprioceptive data to the reticular activating system.

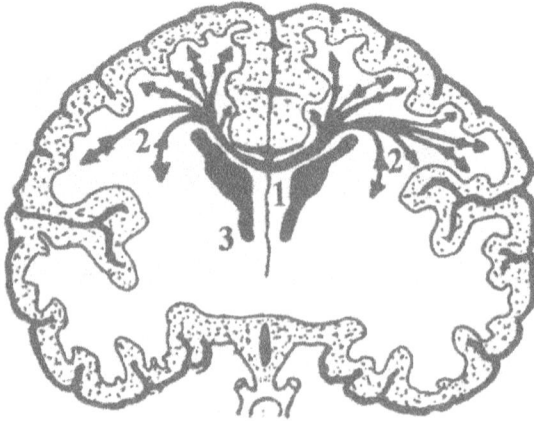

Figure 14

Corpus Callosum: the major tract joining the right and left hemispheres. Sharing information in both directions, the corpus, with the anterior and posterior commissure, is intended to facilitate the maximum simultaneous participation/ interaction within the higher cortex.

1 - Corpus callosum
2- Fibers of the corpus traveling in both directions
3- Lateral ventricles

In summary we cite Emery once again. "But as radiant light does not provide meaningful information and cathode ray technologies are radiant, it becomes difficult to argue that they are providing ecologically or adaptively useful information, and promoting understanding or learning."[100]

Chapter IV

Body Responses –
The Preparation for Action

EACH OF US HAS EXPERIENCED, while watching TV, the increase of heartbeat and breathing rate, the "fidgeting around" of the body at critical moments, perhaps even tears or inner-outer expressions of anger, sudden surprise, or elation. Each of these physical and emotional responses involves the continual manufacture and release of a host of highly refined chemicals (such as adrenaline), major shifts in blood flow through our organs, and complex contractions and moment to moment "resettings" of the tension in our muscles.

This activity is an orchestrated response to images coming from the outside world. It is a very expensive orchestration as it makes use of many of the most refined energies and biochemicals in the body. This orchestration is a preparation for, and sustenance of, action.

The action, however, almost never takes place. We are engulfed in synthetic images that provoke complex biochemical and neuromuscular preparatory or reactive states that are rarely, if ever, fulfilled in a real life "whole person" event. As one researcher put it, "The action may only play itself out after the TV goes off."

Is there any long term physiological significance deriving from these recurring cycles of preparation and support for actions that never take place? We can point to few specifics, but the question has not been comprehensively investigated. The few studies noted here demonstrate an arena of real phenomena inferring a broad range of physiological questions which should be addressed. In 1992 a study of a thousand children (ages 2–20) pointed to a clear relationship between 2–4 hours of daily TV viewing and a dramatic predisposition to elevated cholesterol levels.[101] A study at the University of Tennessee in 1993[102] concluded that "television viewing has a fairly profound lowering effect on metabolic rate and may be a mechanism for the relationship between obesity and amount of television viewing." The decrease in metabolic rate was greater than that found at rest. Several studies have demonstrated that the amount of time spent watching TV (ages 6–11) was a clear predictor of those who would become obese at ages 12–17.

In 1994 Balague[103] reported on a study of 1755 Swiss children 8–16 years of age regarding the relationship between a number of social factors (including TV viewing time) and low back pain. A statistically significant relationship was found between the reporting of low back pain and 2 or more hours (daily) of television viewing in the preceding week. This appeared to be independent of other activities or a parental history of low back pain. A study reported in 1997 on National Public Radio showed dramatic evidence of bronchospasm and increased breathing and cardiac rate when asymptomatic asthmatic children were shown the scene in the movie "E.T." where the "extraterrestrial" apparently died. It was not reported whether this was TV- or movie-viewing. In either case, however, the evocative role of "image" is clear.

We can also point to a number of studies that raise concerns about the organic effects of prolonged or chronic elevations in adrenaline-like substances, and we can ask many common-sense questions about the value and purpose of any unreal and partial "event" which provokes large scale biochemical and physiological reaction. Such questions could include:

Is the growth and development of my child's brain and body assisted by this repetitious flood of biochemicals? Knowing that humans are extremely responsive, emotionally, to facial images, what does it mean to provoke emotions that are disconnected from the real-time life experience of my child? Is their emotional world enriched, and are they thereby enabled to deal more maturely with the stresses of daily life?

Much emphasis has been placed by some on the educational values of television (presumably referring to the education of the neomammalian cortex). Our own experience, in the flow of the life of children-becoming young adults, is quite the opposite. We find, within a short length of time, remnants of "facts" recalled (and those frequently mixed up) but no wholed (body, emotions, mind) experiences that reflect a growth in values, perspectives, or analytical and critical abilities. Recall of educational TV programs, when compared with field trips, lab experiments, or insightfully guided discussions, is factually atrocious, easily swayed, and poorly defended when intellectually challenged. Where are the detailed, scientific studies that could evaluate the physical, emotional, and intellectual responses and the fusions of focus and context that are the real measure of learning and the attainment of intelligence?

Passive or Active?

Many questions inevitably arise concerning cathode ray tube technology in other fields, e.g., video games and computers. People frequently ask, "If the disharmonies you emphasize are operating at pre- and subconscious levels, doesn't that mean that they are also present during all other activities that involve people looking at a cathode ray tube?" The answer is "yes – maybe, and – no." Because the reasons for this seemingly equivocal answer are of singular importance to our aim, they will be taken up individually.

A. A very large proportion of television viewing takes place with the viewer in a physically and mentally passive state. By this we mean to emphasize that under most circumstances one does not sit down to watch TV with an aim to intellectually explore, evaluate, criticize, compare, or come to a fuller understanding of. We sit down to be entertained. We are passive participants, with no recognition of any need to be vigilant, cautious, or alert to small changes. This passivity of the brain sets the stage for all the neurophysiological events that follow.

B. There is no expectation that a responsible action must follow on our viewing. There is nothing to do except keep watching. In fact, there is most often nothing that could be done that relates specifically to the program content.

C. As mentioned earlier, the great majority of events portrayed on the TV screen have no contextual meaning within the immediate timeline of the viewer. Sitting in a comfortable chair in a warm living room, neither hungry nor in pain, the physical, emotional, and intellectual brained activity is wholly out of context with whatever is on the screen. One is not there on the streets of Beirut or the jungles of Burma, nor is one there in the football stadium or as an observer of a murder or as part of an audience at a quiz show. Viewers often comment on the powerful feeling of familiarity with these events, and much of this feeling is related to the prefabrication of the images and the affective reactions provoked in the limbic brain.

D. The viewer's brain does not have to create any visual images (they are totally supplied on the TV screen). The relative uniqueness of this situation in the evolutionary history of the brain of the human being has not as yet been considered from a scientific perspective. From the earliest of times, from storytelling to reading books, the brain constructed its own images, evoked by word or by gesture and tone of voice, either by a person who was real in that moment, or via the active reading of the words.

A nine-year-old child made this emphatically clear during a reading by his schoolteacher of *Robinson Crusoe*. After reading several chapters aloud the teacher came upon a series of pictures portraying the major characters in action. He passed the book around so that all the children could look at the pictures. The nine-year-old boy said, after looking at the pictures, "I like my pictures better!"—and a chorus of "me, too" came from the class. (The children, as a group, had watched very little television at home.)

Adults also reflect this important difference between forming their own images (visualizing) and having them prepared and, in a sense, forced on their brains from outside. A Jewish acquaintance of mine said, only half-joking, that he intensely disliked Charlton Heston because the visual image he had formed over many years, from reading and listening to stories ("his" Moses) was so much more real and meaningful to him and yet, because of the power of the movie image of Charlton Heston portraying Moses, he was unable to visualize his own Moses without getting it mixed up with the movie Moses.

This is not an insignificant point. The images we form ourselves, as we read or listen, have complex and subtle self-meanings that enfold aspects of our physical, emotional, and intellectual life. They are, in a real way, masterful creations that often meld many levels of meaning into one dramatic expression.

The development of the capacity to visualize (or to create other wholes in our sensate world, e.g., hearing music, tasting foods, or recreating an image of the tactile sensations that come when we dive into cool water) does not appear ready-made in the brain. We all have an elemental neural capacity in this regard, but for it to become the valued and enormously useful tool that it can become, the capacity must be exercised, practiced, evoked by parents and teachers, and challenged to become as versatile an ability as possible. Television viewing, regardless of program content, blunts this capacity, because it supplies a steady stream of prefabricated images. The growth and development of this creative capacity to "visualize-by-oneself" is thereby stunted.

E. With respect to video games, the passivity of the participant is of a somewhat different order. By manipulating a few buttons (in neuromuscular terms an extremely simple event), a subjective impression of real participation is created, largely because of the interactive nature of the prefabricated images. The illusion of a real interaction intensifies the limbic brain's conviction that something far more real than cathode tube viewing is going on. This becomes a potent evoker of emotional reactions, reactions that are wholly out of context with the real time life-event of the person.

There is a powerful sense of doing something, of being wholly involved in some meaningful life activity, that is produced by video games. The heart accelerates, breathing increases, activating hormones are released, muscles tense (just as in television viewing), but the entire active involvement is encompassed by pushing buttons! It is also interesting to note just who plays video games. In the Mall video establishments, of which we are aware, we have never seen a mother, and have rarely seen a girl, playing video games. What does that mean? What do these men and boys learn from this highly-charged but only button- pushing activity that emphasizes violence, competition, and speed?

F. Computer usage – air traffic controllers The most obvious and important neurologic difference between television viewing and viewing a computer screen lies in the purposive use made of a computer by the operator. In that respect it is used much like other tools. The user is in command, actively directing the tool and, for the most part, has aims beyond the data, computation, drawings, etc., that are the functional output of the computer. This type of attended computer usage is diametrically opposite to the passive state of television viewing, and the neural processing would, therefore, be different in a number of ways.

Many uses of the computer (including air traffic controllers) also require the operator to break contact with the screen frequently, and to the degree that this occurs, the potential for habituation or neural entrainment is lessened. It is also true that images, while still a dominant feature, are often not as focused on the evocation of limbic (emotional) reactions as is true in television viewing. In spite of this, we must point to the studies reported herein that have shown distinct differences in beta wave brain activity when reading print vs. reading the same text on a cathode ray tube.

We have barely moved into the age of computers, and thorough neurophysiologic studies have not been undertaken as yet. It would be folly to conclude, simply because computers are used and a far higher level of intention-attention may be present in the user, that the other neural events discussed in previous chapters are unimportant, inoperative, or nullified.

Chapter V

Program Content

WE EMPHASIZED AT THE OUTSET that our exploration would not concern the program content of contemporary television. We hope it is evident now that it is essential to have a better understanding of the processes that are occurring in thousandths to hundredths of a second, considerably before the associative cortex of the third (neomammalian) brain can be fully engaged. This plenum of unconscious (i.e., core sensory processing) and subconscious (limbic and neomammalian associative processing) events establishes the neural background (context) on which any program, of whatever content, is focused.

That the specifics of program content can play into the disharmonies already established and further exaggerate and/or distort the physiological reactions is obvious. The peculiar set of vulnerabilities and disharmonies discussed previously will affect human beings in many and different ways varying across a spectrum that reflects the complex mixture of hereditary, familial, and environmental factors that produce relatively unique responses on the part of each human being. Hence, a program with violent or sexually explicit content (as a focused event) will be processed within the context already established by hereditary and environmental factors, as well as by the unconscious and subconscious factors referred to throughout the book. While averages and statistical norms of response may be arrived at via population studies, we wish here to emphasize the individuating factors.

Program content is, thus, a secondary consideration, one that gains the fullest extent of meaning only when the vulnerabilities of the human brain to television viewing have been taken into proper account.

Final Questions

If so many of the human being's makings and creatings are of the nature of prefabricated or synthetic sources of image, how are we to approach them, given the real and potential vulnerabilities discussed in the body of this book? This question has particular application to our children and grandchildren, when we consider that the presence of sources of synthetic images has shown no signs of decreasing its meteoric ascent over the past fifty years. In fact, with the explosive growth of computer usage, and the equally rapid spread of the Internet, the "ocean-of-images" referred to in the Preface will only become larger and more densely packed.

Since the human being's third or neomammalian brain has been the source of this image-making, it would seem that the answers must come from here as well. The answers will require the best of our intelligence (in the sense of the harmonious fusion of focus and context referred to earlier).

It seems quite clear that exposing the core and limbic brains to an unfettered onslaught of prefabricated images has been a great folly. One need only follow a mother with children through a supermarket and notice the totality of non-evaluative "I want," "I need," "Get that!" that streams from the children in an unthinking reaction to advertising—or simply watch any of hundreds of clever TV or magazine ads that are focused almost exclusively on the provocation of core (sensorimotor) and limbic reactions. The third brain is an absent or minimal participant in this plethora of daily events. It reads and says the words, hears the vocalization of pleasure, but analyzes, evaluates, criticizes, and discriminates next to nothing.

It is remarkably (and sadly) similar in the adult world. From cigarette, lingerie, and beer ads to political speeches, social commentary, and religious tirades, the sensitivity and temporal vulnerability of the core and limbic brains are plumbed by evermore and self-serving synthetic images. Given this circumstance, is it possible to protect and guide the essential innocence of a growing brain, a brain that accepts the world out there and in here as the truth, because it is created to respond that way?

If attention-vigilance mechanisms are severely inhibited; if the critical-analytical powers of the left cortex are dampened, while the sensitive interconnection between the image-vulnerable right cortex and the limbic brain is left intact; if midbrain searching mechanisms can be habituated and/or entrained; if the effector function of radiant light is an operant part of television viewing; if focus and context are never reconciled or fused into a meaningful whole—if any or all of these changes are intrinsic to the neurophysiologic process of television viewing, then we have set in motion an environmental hazard to normal brain growth and development of unmeasured dimensions.

Is it possible—is it a present day individual and societal responsibility—to enable and train the neomammalian brain, from the first education by parents all the way through collegiate and professional education, to be more alert, vigilant, and discriminating of the images its senses must accept?

Given the questions and preliminary findings incorporated in this short book, and given the massive economic, political, and entertainment dependencies on the television industry that have developed during the past fifty years, is it possible to look at the difficulties with impartiality and reason?

Could we make such an effort for our grandchildren?

ENDNOTES

1 Walsh, Roger. "Effects of Environmental Complexity and Deprivation on Brain Chemistry and Physiology: A Review," *International Journal of Neuroscience*, 1980, p.77.

2 Libet, Benjamin, et al. "Time of Conscious Intention to Act in Relation to Onset of Cerebral Activity [Readiness - Potential]," *Brain*, 1983, 106, pp. 623–42.

3 Nauta, W. and Feirtag, M. "The Organization of the Brain," *Scientific American*, September 1979, p.88.

4 Bohm, David. *Wholeness and the Implicate Order*, London: Arc Paperbacks, 1980, p. 3.

5 *Nielson Report 1997*, Nielson Media Research.

6 Ibid.

7 Mulholland, Thomas B. "Training Visual Attention," *Academic Therapy*, 1974, 10.1. pp. 5–17.

8 Ganong, W.F. *Review of Medical Physiology*, 1977, pp. 88–99.

9 Watts, George. *Dynamic Neuroscience*, Harper and Row, 1975, p.45.

10 Chusid, J.G. *Correlative Neuroanatomy and Functional Neurology*, 16th ed., Lange Publishing, 1976, pp. 91–112.

11 Best & Taylor. *Physiologic Basis of Medical Practice*, 11th ed., Baltimore: Williams and Wilkins, 1984

12 Guyton, A. *Textbook of Medical Physiology*, W.B. Saunders, Co., 1981, Chs. 54 and 55.

13 Ibid. Ch. 55, pp. 684–690.

14 Emery, F.E. and Emery, M. *A Choice of Futures: To Enlighten, or Inform*, 1975, Center for Continuing Education, Australian National Univ., Leiden: Martinies Nijhoff Social Sciences Division, 1976, pp. 75–81.

15 Guyton, A. *Textbook of Medical Physiology*, Ch. 55.

16 Schnapf, J. and Baylor, D. "How Photoreceptor Cells Respond to Light," *Scientific American*, April 1987, pp. 40–47.

17 Finke, R.A. "Mental Imagery and the Visual System," *Scientific American*, March 1986, pp. 88–95.

18 Treisman, A. "Features and Objects in Visual Processing," *Scientific American*, Nov. 1986, pp. 114–125.

19 Gibson, J.J. *The Ecological Approach to Visual Perception*, Boston: Houghton-Mifflin, pp. 246–250.

20 Wilson, A.D. "The Molecular Basis of Evolution," *Scientific American*, Oct. 1985, pp. 1709–1773.

21 Cann, R.L., Stoneking, M. and Wilson, A.C. "Mitochondrial DNA," *Nature*, 325:31, 1987.

22 Wainscoat, J. "Mitochondrial DNA Evolutionary Tree," *Nature*, 325:31, 1987.

23 Gibson. 1979, pp. 244–245.

24 Ibid., pp. 245–246.

25 Brou, P. et al. "The Colors of Things," *Scientific American*, Sept. 1986, pp. 84–91.

26 Gibson, 1979, p. 242.

27 Wurtman, R. "The Effects of Light on the Human Body," *Scientific American*, July 1975, pp. 69–77.

28 Smith, Kendric. "The Science of Photobiology," *Bioscience*, Vol. 24, Jan. 1974, pp. 46–47.

29 Rosenthal, N.E. et al. "Antidepressant Effects of Light in Seasonal Affective Disorder," *American Journal of Psychiatry*, 142:2, Feb. 1985.

30   Smith, K. 1974, p. 74.
31   Borbely, A. "Effects of Light on Sleep and Activity Rhythms," *Progress in Neurobiology*, Pergamon Press, Vol. 10, 1978, pp. 1–31.
32   Emery, Merrelyn. "The Social and Neurophysiological Effects of Television and Their Implications for Marketing Practice: An Investigation of Adaption to the Cathode Ray Tube," Australian National University, Melbourne, 1985.
33   Emery, F. and Emery, M., 1975.
34   Libet et al. 1983.
35   Gibson. 1979.
36   Pawley, Martin. *The Private Future*, 1979, London: Thames & Hudson.
37   Emery, M. 1985, p. 22.
38   Geschwind, N. "Specializations of the Human Brain," *Scientific American*, Sept. 1979, pp. 180, 190–192.
39   Emery and Emery. 1975, p. 89.
40   MacLean, Paul D. "Psychosomatic Disease and the Visceral Brain," *Basic Readings in Neuropsychology*, Harper and Row, 1964, p. 185.
41   Guyton. 1981, p. 677.
42   Ibid., pp. 675–676.
43   Phelps, N.E. et al. "Metabolic Mapping of the Brain's Response to Visual Stimulation: Studies in Man," *Science*, 1981, (a) 211.1445 8.
44   Engel, Jerome. "Discussion: Positron Imaging of the Normal Brain-Regional Patterns of Cerebral Blood Flow and Metabolism," *Transactions of the American Neurological Association*, 1980, 105. pp. 9–10.
45   Nielsen. 1987, p. 4.

46　Krugman, Herbert E. "Brainwave Measures of Media Involvement," *Journal of Advertising Research*, 1971, 11:1 pp. 3–9.

47　Bogart, L. et al. "What One Little Ad Can Do," *Journal of Advertising Research*, Vol. 10 #4, Aug. 1970, pp. 3–15.

48　Krugman, H. "The Measurement of Advertising Involvement," *Public Opinion Quarterly*, Winter 1966–1967, pp. 583–596.

49　Thomas, E. " Movements of the Eye," *Scientific American*, Aug. 1968, pp. 88–95.

50　Mackworth, N.H. "A Stand Camera for Line-of-Sight Recording," *Perception and Psychophysics*, March 1967.

51　Krugman. 1971, p. 4.

52　Ibid., p. 4.

53　Ibid., p. 8.

54　Rossiter, John. "Point of View: Brain Hemisphere Activity," *Journal of Advertising Research*, 1980, 20.5, pp. 75–76.

55　Beatty, J. and Wagoner, B. "Pupillometric Signs of Brain Activation Vary with Level of Cognitive Processing" *Science*, Vol. 199, 1978, pp. 1216–1218.

56　Featherman, G. et al. "Electroencephalographic and Electro-oculographic Correlates of Television Viewing," National Science Foundation, March 1979, final technical report.

57　Krugman. 1971, p.8.

58　Appel, V. et al. "Brain Activity and Recall of TV Advertising," *Journal of Advertising Research*, 1979, 19.4, pp. 7–14.

59　Silberstein, R. et al. "Electroencephalographic Responses of Children to Television," Australian Broadcasting Tribunal, Melbourne, 1983.

60   Mulholland. 1974, p. 8.

61   Ibid., p. 14.

62   Mander, J. *Four Arguments for the Elimination of Television*, New York: Quill, 1978, p. 209.

63   Emery, M. 1985, p. 627.

64   van Lith, G.H. et al. "Two Disadvantages of a Television System as Pattern Stimulator for Evoked Potential," *Doctor of Ophthalmology*, April 15, 1980, 48 (2), pp. 261–266.

65   _____. "Interference of 50 Hz Electrical Cortical Potentials Evoked by TV Systems," *Doctor of Ophthalmology*, Nov. 1979, 63 (11), pp. 779–781.

66   Ackoff, R. and Emery, F. *On Purposeful Systems*, 1972, Aldine-Atherton or Tavistock, Intersystems, Inc. 1981.

67   Krugman, H. 1971, pp. 8–9.

68   Cooper, R. et al. "Regional Control of Cerebral Vascular Reactivity and Oxygen Supply in Man," *Brain Research*, 1963, Vol. 3, pp. 174–191.

69   Emery, M. 1985, p. 639.

70   Phelps, M.E. 1981.

71   Enge, U. 1980.

72   Simon, O. et al. "The Definition of Waking Stages on the Basis of Continuous Polygraphic Recordings in Normal Subjects," *Electroencephalogical Clinical Neurophysiology*, 1977, # 42, pp. 48–56.

73   Emery, M. 1985.

74   Winn, Marie. *The Plug-in Drug*, New York: Viking, 1977.

75   Mulholland. 1974.

76   van Lith et al. 1980.

77   New Scientist, 1983, p. 623.

78   MacLean. 1964.

79   Lange, L.S. "Television Epilepsy," *Electroencephalogical Clinical Neurophysiology*, 1961, #13, p. 490.

80 Charlton, M.H. and Hoefer, P.F.A. "Television and Epilepsy," *Archives of Neurology*, #11, pp. 239–247.

81 Jeavons, P.M. and Harding, G.F.A. (eds). "Photosensitive Epilepsy," *Clinics in Developmental Medicine*, No. 56, London: W. Heineman Medical Books, Ltd., 1975.

82 Wilkins, A.J. et al. "Visually Induced Seizures," *Progressive Neurobiology*, 15 (2), pp. 85–117.

83 Daneshmend, T.K. et al. "Dark Warrior Epilepsy," *British Medical Journal*, 1982, 284, 6331, pp. 1751–1752.

84 Emery. 1985, p. 659.

85 Klapetek, J. "Photogenic Epileptic Seizures Provoked by Television," *EEC and Clinical Neurophysiology*, 1959, #11, p. 809.

86 Anderman, F. "Self-Induced Television Epilepsy," *Epilepsia*, 1971, #12, pp. 269–275.

87 Bickford, R.G. and Klass, D.W. "Stimulus Factors in the Mechanism of Television Induced Seizures," *American Neurological Society*, 1962, pp. 87–176.

88 Wilkins, A.J. et al. 1980.

89 Ibid., p. 86.

90 Wilkins, A.J. "Television-Induced Epilepsy and Its Prevention," *Br. Bed. Journal*, 1978, 31, pp. 1301–1302.

91 Johnson, L.C. "Flicker as a Helicopter Pilot Problem," *Aerospace Medicine*, 1963, #34, p. 306.

92 Jeavons, P.M. 1975.

93 Darby, C.E. et al. "The Self Induction of Epileptic Seizures by Eye Closure," *Epilepsia*, 1989, #21, pp. 31–42.

94 Korr, I.M. "The Neuromusculoskeletal System as the Instrument of Life," *Journal of the American Academy of Osteopathy*, October 1979.

95   Frymann, V.N. "The Law of Mind, Matter and Motion," Scott Memorial Lecture, *Journal of the American Academy of Osteopathy*, pp. 57–66.

96   Korr, I.M., Buzzell, K.A. and Hix, Elliott. "The Physiological Basis of Osteopathic Medicine," NY Post Graduate Institute of Osteopathic Medicine and Surgery, 1970.

97   Dunn, Gwen. *The Box in the Corner*, London: MacMillan, 1977.

98   Emery, M. 1985, p. 628.

99   Emery and Emery. 1975, p. 79.

100 Ibid.

101 Wong and Hei. "Television Viewing and Pediatric Hypercholesterolemia," *Pediatrics*, Vol. 90, #1, July 1992.

102 Klesges and Shelton. "Effects of Television on Metabolic Rate: Potential Implications for Childhood Obesity," *Pediatrics*, Vol. 91, #2, Feb. 1993.

103 Balague, F. et al. "Non-specific Low Back Pain among School Children," *Journal of Spinal Disorders*, Vol. 1, #5, pp. 374–379, 1994.

www.ingramcontent.com/pod-product-compliance
Lightning Source LLC
Chambersburg PA
CBHW022340280326
41934CB00006B/713